T0222455

The Star and the Whole

The Star and the Whole

Gian-Carlo Rota on Mathematics and Phenomenology

Fabrizio Palombi

Translated by Giacomo Donis

CRC Press
Taylor & Francis Group
Boca Raton London New York

CRC Press is an imprint of the
Taylor & Francis Group, an **informa** business
AN A K PETERS BOOK

CRC Press
Taylor & Francis Group
6000 Broken Sound Parkway NW, Suite 300
Boca Raton, FL 33487-2742

© 2011 by Taylor & Francis Group, LLC
CRC Press is an imprint of Taylor & Francis Group, an Informa business

No claim to original U.S. Government works

Printed in the United States of America on acid-free paper
Version Date: 20110714

International Standard Book Number: 978-1-56881-583-1 (Paperback)

Library of Congress Cataloging-in-

Palombi, Fabrizio, 1965-
 The star and the whole : Gian-Carlo Rota on mathematics ar
translated by Giacomo Donis.
  p. cm.
 "Published with the contribution of the Faculty of Letters ar
Calabria."
 "An A.K. Peters book."
 Includes bibliographical references and index.
 ISBN 978-1-56881-583-1 (pbk.)
 1. Rota, Gian-Carlo, 1932-1999. 2. Mathematics--Philosoph

QA29.R68P3513 2003
510.1--dc23

Visit the Taylor & Francis Web site at
http://www.taylorandfrancis.com

and the CRC Press Web site at
http://www.crcpress.com

K00789_Discl.indd 1

To the memory of Gian-Carlo Rota,
dearest master and friend.

Table of Contents

Acknowledgments

I wish to thank Giacomo Donis for his passionate and excellent work of translation, and Bruna Mancini for her precious revision of the proofs. My reconstruction of Rota's human and intellectual events made use of the help of Ester Rota Gasperoni, Ottavio D'Antona, and Mark Van Atten. I also thank Francesca Bonicalzi and Gianfranco Dalmasso for their important advice and encouragement.

The publication of this book would have been impossible without the precious and generous support of Henry Crapo.

Foreword

This admirable book by Fabrizio Palombi is an attempt to "figure out" the life, work, and person of Gian-Carlo Rota. The English word *figure* enjoys multiple meanings, and all of them bring out different facets of Rota as the distinguished subject of this book. As a noun, *figure* can mean the shape of a thing, but it can also mean the numerals (figures) used in mathematics. It can signify a personage or a character, an imposing figure. The verb *to figure* can mean to calculate, or to portray in pictures, sculptures, or diagrams. The verb *to figure out*, however, means to resolve or understand something perplexing, and, in another sense, it can signify to fill out or to present in a concrete, embodied manner.

When something striking happens it takes time to know—to figure out—what it was. Aristotle has a wonderful phrase to express the essence of a thing. He calls it "the what it was for this thing to be." The phrase uses the past tense. Only when a thing has receded some distance in time from us can we begin to see "what it was for it to be." Only then do we start to see it in perspective and sift out its essentials. People who argued with Socrates were dazzled, but Plato showed what it was for Socrates to be. Plato could look back and see more clearly what was going on. Rota was remarkable in this way, as a man and as a thinker, and we are indebted to Fabrizio Palombi for sifting through Rota's memories and writings to help us understand him and his work.

The book has four sections. The first recalls the salient events in Rota's life, while the next three explore his philosophy of mathematics and human cognition. Rota's authority as a thinker is based on his mathematical accomplishments; not just because his reputation as a mathematical genius made people pay attention to his philosophy, but because his actual activity and insights in mathematics were what he thought about. He did not just look at the results—the documented proofs and theorems—of other people's work; he thought about what he himself was doing and what he actually managed to see. He was both a mathematician and a philosopher and his life in mathematics gave him the material for his philosophy. To use

Edmund Husserl's phrase, Rota was able to get "to the things themselves" in mathematics because he played a strategic role in their development.

One of Rota's main philosophical complaints is that people usually overlook what is there before them; they oversimplify things and detach them from their contexts. They take a part for the whole. In mathematics, such misplaced concreteness would involve looking primarily at a proof of a theorem as though it were the only thing that mattered in arriving at the truth of the theorem. Rota insists that a lot more is going on that is relevant to what mathematics is. Mathematicians have motivations, beauty and value occur in mathematics, false moves and dead ends along the way have a role to play in mathematical success, and a given theorem may be provable in more than one manner. Such things are not merely psychological or anecdotal accessories to mathematics; they are not decorations or idiosyncrasies. They have a structural role in mathematical achievement and in bringing about— constituting—the objectivity proper to mathematical entities, the "things" that are discovered and that can be handed on from one person to another and from one generation to the next. We need to look not just at mathematical objects but also at the intentionality that both establishes them and brings them to light. More is there than meets the eye, more is heard than is actually spoken. Philosophy needs to explore or at least point to these shadows, absences, and perspectives, which we miss when we focus only on the object and become lost in what Rota calls "objectivity."

Rota defined his work as an exercise in phenomenology and he was inspired by Husserl's philosophy. Husserl's first book was *Philosophy of Arithmetic*, published in 1891. His second was *Logical Investigations*, which appeared a decade later in 1900–1901. The two books set up the polarity within which phenomenology developed. The first shows how groups and numbers are constituted, the second shows how judgments and categorial objects are established. The first sets the philosophical stage for mathematics, the second for logic. Both together open a space for metaphysics or first philosophy in the modern world, in a manner that encompasses not just things but also the subject or the person who is the dative of manifestation for things. Rota's interest in phenomenology led him to emphasize the first of these two themes, the one explored in *Philosophy of Arithmetic*, but he also drew on the second and he was always alert to the human agent engaged in mathematics. One of Jacques Derrida's early writings was a commentary on the beginning of *Logical Investigations*, entitled *La voix et le phénomène* (Derrida, 1967b). It is regrettable that he did not also deal with the book on arithmetic, which might have provided ballast for Derrida's reading of Husserl. Palombi's brief remarks about Derrida and writers such as Lacan open up the project of comparing Rota's thought with that of other contemporary continental thinkers.

Husserl's two books, *Philosophy of Arithmetic* and *Logical Investigations*, treat, respectively, two kinds of intellectual activity: counting and predicating, or reckoning and describing. In the first we assemble items into groups or wholes. In the second we begin with a whole and sort out parts within it; we articulate features or moments within the thing. Counting yields equations, predicating is done through sentences. The two kinds of intelligence are irreducible one to the other, and taken together they make up the core of human thinking. Each of them involves a distinction between content and the operational forms that pattern it.

One major difference between them, however, is that in counting or reckoning we have a much greater control over the operations that we perform. We can devise new ways of grouping items into wholes; we can more consciously focus on the operations and contrive new ones. Instead of just adding units, for example, we can multiply them, and instead of just subtracting we can divide. In predicating, however, we seem coerced into the forms of logic that are given to us by our native language and by the human mind with its conversational possibilities and deep syntax. We don't seem to have much choice in the matter; there is a subject, there is a predicate, and we can't help but start with them and stay with them. It seems that we are not able to abbreviate complicated and multiple predications into simpler ones, the way we can abbreviate complicated mathematical operations into tidier ones. For that reason, mathematics and the sciences that involve mathematics can become extremely exotic and understandable only to the few who have mastered the new operations and their syntax, while speech and its predications cannot depart all that far from ordinary discourse, at least not in regard to logical form. The content is another matter, of course, and poets and novelists might create new kinds of higher-level literary forms, but the basic logical operations inevitably remain what they are.

Palombi discusses the theme of "erasure": the fact that we cover over or forget some of the things we have seen and done as we move toward the intellectual resolution of a problem. The problem once solved, the way we got there is almost forgotten; we fasten on the theorem and "erase" (a) the errors and dead-ends we ran into and have rejected on the way, (b) the valid but complex proofs and procedures that have now been simplified and whose earlier forms have therefore become stale, and (c) the deep intentional strata that make any kind of calculation possible, such as the very possibility of counting, of writing, or of referring to the same thing in different contexts. These things fall into what Husserl calls a "sedimented" condition, forgotten but still effective, still watching and waiting as we are busy with other things. As Husserl, Rota, and Palombi remind us, we need such oblivion if we are to move on toward further resolutions; we cannot turn toward other issues if our hands are full with the past. But we should not delete the things we currently overlook; the formulas we are satisfied with now are not the only things we should appreciate. The intellectual history of mathematics provides resources for further mathematical life.

How do we ourselves fit in with science and its mathematics? Is mathematical science something *we* do, or does *it* tell us what we are and where we came from?

Whose excellence is primary? Does artificial intelligence tell us what natural intelligence is, or is it the other way around? Gian-Carlo Rota saw mathematics as one aspect of human veracity, the inclination toward truth and truthfulness that defines us as human beings. He recognized it as a particularly powerful form of this human capacity; it has made possible the kind of life we are now engaged in, for better or worse. Buildings would not stand nor airplanes fly without it. Rota saw it, however, as a part of human being and not the whole, and through this presentation by Fabrizio Palombi, Rota reminds us of other forms of thinking and living that we dare not forget.

—Robert Sokolowski
Washington, DC
May 2011

Gian-Carlo Rota's Third Way

"The scientific method is not a technique. If it becomes one it betrays its own essence."

—Martin Heidegger

"A star-shape built out of star-shapes, which in their turn are composed of stretches and ultimately of points. The points serve to 'found' stretches, the stretches serve to 'found,' as new aesthetic unities, the individual stars, and these in their turn serve to 'found' the star-pattern, as the highest unity in the given case."

—Edmund Husserl

Gian-Carlo Rota's life was an intense intellectual adventure dominated by an inexhaustible curiosity, as his dozens of publications across numerous fields of mathematics and philosophy attest. This book, the first to examine his philosophical investigation, aims less at historically reconstructing the evolution of his thought than at tracing a common thread, an element of continuity in his heterogeneous interests. This thread is constituted by *Fundierung*, a phenomenological theme espoused by Husserl, which Rota characterized as ranking among Husserl's greatest logical discoveries, and on which Rota focused his attention from the first of his philosophical writings. In Husserl, *Fundierung* is translated as "foundation"; Rota describes it as "layering, letting, and founding" (Rota, 1973a; Rota, 1986a, p. 171). Rota uses *Fundierung* as an instrument to open up a phenomenological way capable of overcoming the classical alternative between empiricism and rationalism that not only conditions philosophical tradition but, in particular, represents an aporetic obstacle for a comprehension of the nature of mathematical entities. In this way Rota speaks for a philosophical project that, from Kant to Bachelard,[1] has sought to overcome the classical forms of the hierarchization of reality stemming from this metaphysical alternative that claims to interpret physical objects as "more real"

---

[1] In this regard see Bonicalzi (1982), pp. 36–39. In particular the schema of note 14 on p. 39 represents the correspondence/opposition of a series of pairs (idealism, conventionalism, and formalism on one side; positivism, empiricism, and realism on the other) that will be used extensively in my own investigation. The individuation of a *middle* philosophical way—the inspiration for my definition of "third way"—is the focus of Bachelard (1972), pp. 27–34. Francesca Bonicalzi has provided many suggestions, contributions, and comparisons that have been of great value for my research.

than objects that are ideal (or vice versa), or to attribute some type of physical existence to every entity.

The common thread of *Fundierung* influences not only the philosophy of mathematics but also the other principal fields of Rota's philosophical investigation: his reflection on objectivism, his rereading of Heideggerian hermeneutics, his critique of analytic philosophy, and his critique of scientism. Objectivism was a crucial issue for many contemporary philosophers—Husserl in particular. But one of Rota's most original contributions was his attempt to single out objectivism as an obstacle not only for philosophical research but also (and above all) for scientific research. His objection to objectivism does not consist in a critique of science *tout court*, but rather of reductionism entangled in the most classical philosophical aporias by its inability to grasp the complexity of the relation between the whole and its parts. This is a question of great topical interest, as was demonstrated by the international conference held at the Università Statale in Milan in 2009, on the occasion of the tenth anniversary of Rota's death.[2]

For Rota, these difficulties can be overcome if, through the relation of *Fundierung*, the full importance of the autonomy of a phenomenon's levels of description can be grasped. His reflections in this regard are contained in his most important text, significantly titled *The End of Objectivity*, which collects the lessons of his philosophy courses at MIT (Rota, 1991a). [3] In this vein, the title of my book is influenced by *Fundierung*; Husserl's "star-shape" in the *Third Logical Investigation* is used to exemplify a particular type of foundation (*Fundierung*) between the whole and the parts that Rota transformed into one of the keystones of his thinking.

The theme of *Fundierung* also conditions Rota's rereading of Heidegger, particularly regarding the problem of "sense" (*Sinn*), its contextual nature, and its connection with a physical substrate. This perspective underlies his philosophical project for a pluralistic cultural and scientific attitude where, alongside non-dogmatic objectivism, phenomenology can carve out a space of its own, creating a fruitful dialogue between scientific enterprise and philosophical reflection.

This type of approach, like that of other scholars of continental origin, met with fierce opposition on the philosophical scene in the United States. Rota responded with a wealth of essays and articles in which he voices an equally vigorous criticism of analytical philosophy. This polemic, moreover, interacted with his interest in issues relative to teaching, to scientific education, and to diffusion, turning his philosophical works into an important reference point for all those who are interested in understanding and overcoming the fracture between analytic and continental philosophy.[4]

---

[2] See D'Antona, Damiani, Marra, Palombi (Eds.), (2009).

[3] This is a 457-page compendium of heavily edited lecture notes from Rota's annual courses at MIT between 1974 and 1991. See also Rota (2007).

[4] Regarding this fracture see D'Agostini (1997), pp. 123–166, and Picardi (1999).

The Star and the Whole stems from a debate that I have not resigned myself to consider closed by Gian-Carlo Rota's death. Its path, studded with thoughts and second thoughts, represents the outcome of my interminable and impassioned discussions with my master. This complex structure particularly characterizes the part dedicated to the philosophy of mathematics, in which I propose a deeper reading of certain cues in Rota's thinking based on my own specific interests.[5] This has produced "short circuits" among different philosophical traditions that show (beyond the obvious differences) some profound convergences. In particular, I have attempted to rethink some notions in the thought of Edmund Husserl, Martin Heidegger, and Jacques Derrida in an epistemological vein and to relate them to the thinking of other authors who have specifically analyzed the scientific enterprise, such as Pierre Duhem, Ernst Mach, Henri Poincaré, and Imre Lakatos. This operation is risky from the historical perspective, but is also theoretically interesting precisely because it makes it possible to fix and filter those moments of personal contact with Rota that would otherwise be lost.

The book consists of four chapters and a bibliographical section. The first chapter reconstructs the fundamental coordinates of Rota's cultural biography and examines his peculiar philosophical style, his criticisms of analytical philosophy (against the backdrop of a contraposition that left its mark in US culture), and his reflection on Heidegger's thought.

The second chapter presents a general picture of Rota's personal reinterpretation of phenomenology, influenced by theoretical and didactic demands, focusing in particular on the Heideggerian themes of context and tool. In this chapter the star-shape becomes a powerful instrument for understanding the properties of Husserl's mereology[6] and for the critique of objectivism.

The third chapter proposes an analysis of that which Rota described as the double life of mathematics. This chapter consists of a theoretical reflection on the nature of mathematical entities, which relates Rota's phenomenological investigation—attentive to their constitutive and historical aspects—to his epistemological considerations.

The fourth chapter critically examines the complex relation of mathematical research with technological applicability and scientific progress, based on a phenomenological reflection on intentionality.

---

[5] There are traces of this long debate not only in my memory but also in some 25 audiocassettes that recorded some of our conversations from 1990 to 1998. The present text comprises, moreover, Palombi (1997), (1999a), (1999b), and (2005).

[6] "Mereology (from the Greek μερος, 'part') is the theory of parthood relations: of the relations of part to whole and the relations of part to part within a whole. Its roots can be traced back to the early days of philosophy [...]. As a formal theory of parthood relations, however, mereology made its way into our times mainly through the work of Franz Brentano and of his pupils, especially Husserl's third *Logical Investigation*" (Varzi, 2004).

The text concludes with a full bibliographical section that has a value all its own, capable of tracing the general perimeter of Rota's investigation, of drawing a map of his philosophical interests, and of surveying his sources and his authors of reference. This, to date, is the most complete bibliographical analysis of his philosophical writings, made possible (despite the scattering of his archives) by the notable quantity of material, documents, and books that, in ten years of collaboration, Rota entrusted to me with an explicit request for me to study, catalogue, criticize, and publicize it. I hope that this book may be a fulfillment—however partial—of his wishes.

CHAPTER ONE

✧ ✧ ✧ ✧

Rota, Philosopher

"Our method of presentation is that of phenomenological description. We do not claim to give a 'pure description of the things themselves,' as Husserl wanted. Phenomenological descriptions, far from being pure, are always motivated."

—Gian-Carlo Rota

I believe that the reading and the analysis of a scholar's texts represent the fundamental perspective under which his or her research is to be judged. Likewise, I am convinced that such reading and analysis are endowed with a fundamental degree of freedom with respect to the facts of that scholar's life. I subscribe to the proverbial analogy that equates the relationship between a writing and its author to that between a son and his father—with the passing of time, both acquire increasingly greater independence.

That being said, I think that biographical research can contribute to understanding the genesis of the ideas of a scholar such as Rota, whose work ranged through spheres as significant as they were heterogeneous. For this reason, in the effort to draw a picture of his philosophical work it is important to begin by saying something about his cultural background and development. Obviously, the remarks on specific sectors of Rota's scientific activity make no claim to provide an exhaustive analysis but serve only to provide the reader with some cultural coordinates. In light of the vastness of his scientific and cultural engagement[1] I shall limit myself to a selection of the most important testaments to his intellectual adventure. The first section of this chapter is based on some autobiographical writings of Rota's,[2] on the testimony of relatives and friends,[3] and on my personal recollections.

---

[1] The c.v. he drew up in 1996 lists over a hundred academic degrees and academic and editorial appointments; see Rota (1996).

[2] Rota (1985a), (1987b), (1988a), (1997b), and (1997c).

[3] In particular Rota Flaiano (1997a) and (1997b), Rota Gasperoni (1995) and (1996), Cerasoli (1999), D'Antona (1999a), and Senato (1999).

1.1 ✧ Biographical Notes

Gian-Carlo Rota was born in Vigevano, a town in Lombardy near the Piedmont border, on April 27, 1932, into a family of great cultural traditions. His father, Giovanni (1899–1969), was a well-known engineer and architect,[4] a professor at the University of Quito, and a highly cultured scholar, who made excellence the permanent and almost obsessive goal for him and for his children. His aunt, Rosetta Rota (1911–2003), was a mathematician associated with the renowned Rome University Institute of Physics in Via Panisperna, and married the celebrated writer and screenwriter Ennio Flaiano (1910–1972). These three great figures left an early mark on young Gian-Carlo's life and education, as did his readings in his father's huge private library, which included texts not only on engineering and architecture, but also on mathematics, art, literature,[5] and philosophy.[6] It was here that Gian-Carlo made his first contact with philosophy—through the works of Benedetto Croce in particular.[7]

Rosetta Rota was a brilliant scholar. Studying in Rome with teachers of the caliber of Vito Volterra[i] and Enrico Fermi,[ii] she earned degrees in mathematics in 1929 and in statistical sciences in 1930 and, subsequently, worked as an assistant at Rome University and at the Institute for Applied Calculus. Through this Institute, which dealt with the numerical aspects of the phenomena studied by the Via Panisperna physicists, Rosetta cultivated professional relationships and friendships[8] with Fermi, Franco Rasetti,[iii] Bruno Pontecorvo,[iv] and Emilio Segrè.[v] The life stories of this group's members followed a tragic script, similar to that of so many other great European intellectuals. Here, the intellectual suicide of the Old Continent was played out; political, cultural, and racial persecutions, culminating in World War II, drove these scientists to seek refuge in North America, where they contributed to the creation of that Golden Age of United States science described by Gian-Carlo Rota in a famous essay (Rota, 1988a).

In Rome Rosetta met Ennio Flaiano (whom she married in 1940), a man who represented a constant source of inspiration for Gian-Carlo Rota. In fact I recall

---

[4] Giovanni Rota was an innovator particularly in the field of structures in reinforced concrete (the trusses of the URSUS factory in Vigevano provide an interesting example), and in the earthquake-proof structures built in Quito, Ecuador.

[5] Unfortunately, Giovanni Rota's archives have been lost, while the books of his library regarding architecture, furnishings, city planning, and engineering were deposited in the Historical Library of the Vigevano Engineering Department. The fact that nearly all of the 500 books in the Vigevano Historical Library come from Rota's private collection gives some idea of the enormous size of his library. See Dulio (2000), pp. 10–12.

[6] Both Gian-Carlo Rota and Rosetta Flaiano showed me on several occasions, in Cambridge and in Lugano, a collection of philosophy books published by Laterza that had belonged to Giovanni Rota.

[7] Gian-Carlo Rota, conversation with the author.

[8] See Rota Flaiano (1997b).

that, from the day of our first long conversation in Rome in the summer of 1990, I was struck by how frequently Rota mentioned his name. I subsequently understood that Rota still found in Flaiano, so many years after his death, an intellectual point of reference and a model for his own writing. Rosetta, due to the illness of her daughter Luisa,[9] gave up her work for a long time, taking it up again from 1964 to 1976 at the Universities of Rome and Naples.

Gian-Carlo Rota began his studies in his native Vigevano[10] where, from 1939 to 1945, he attended middle school quite irregularly, on account of the war and of family vicissitudes. Most dramatically, his father was arrested in 1943 as a result of his fierce anti-fascism, but was liberated almost immediately by a sort of popular uprising.[11] After the Armistice of Cassibile on September 8th the senior Rota had to take refuge in Switzerland for a long period of time, where the hardship of exile was aggravated by his difficulties in getting news of his family. Intransigently democratic and secular, anti-fascist but also sharply critical of the methods of political struggle of the Resistance, Giovanni Rota was in contact with the anti-fascist organization Giustizia e libertà (Justice and Liberty),[12] probably through Flaiano's mediation. The risk he ran is testified to by the publication of a leaflet, on January 24, 1944, in which his name appears at the top of the list of wanted subversives, together with Luciano Mastronardi (father of Lucio, the author of *Il maestro di Vigevano*),[13] with whom he had become friends (Zimonti, 1983, p. 144).[14]

In 1944 and 1945, after his father had found refuge in Switzerland, Gian-Carlo together with his mother Luigia Facoetti Rota (1906–1991) and his sister Ester moved into a villa on Lake Orta that belonged to his mother's family (Rota, 1998e, p. 257) During the war his family compensated for the problems with formal schooling by teaching him Latin, French, and mathematics privately. Although Rota did not speak willingly about the events of that period and often referred those who were interested to his sister's memoirs,[15] he did tell me some picturesque anecdotes about the adversities of those times. In particular he was proud of the daily crossing of Lake Orta (in a rowboat hoisting a white flag) that he undertook courageously to get to school. This ritual was the fruit of a complex agreement between fascists and partisans to avoid exposing the group

---

[9] See Guerra (1994). This remembrance of Ennio Flaiano and of his daughter is contained in the anthology *Mi riguarda* (Bossi Fredigotti et al., Eds., 1994), whose publication was promoted by the Luisa Flaiano Foundation created by Rosetta Flaiano in memory of her daughter, who died on January 3, 1992.

[10] One of his schoolmates was the writer Lucio Mastronardi (1930–1979).

[11] Rota family members; conversations with the author.

[12] See Dulio (2000), p. 13.

[13] Luciano Mastronardi's wife, Maria Pistoia, had been Rosetta's schoolteacher; see Rota Flaiano (1997a).

[14] Gian-Carlo was so proud of his father's past that in the living room of his Cambridge apartment he hung a reproduction of the notorious "black list" leaflet.

[15] See Rota (1998e), p. 257.

of children—Gian–Carlo Rota included—to the crossfire of the warring parties that faced off on the opposite shores of the lake.[16]

In 1947, at 15 years old, after finishing middle school, Gian-Carlo Rota followed his family to Quito, Ecuador, where his father had moved to pursue his professional activity and out of fear that post-war tensions could lead to a new war with Yugoslavia (Cerasoli, 1999, p. 16, and Dulio, 2000, p. 14). Giovanni Rota considered the United States to be, unquestionably, the worldwide reference point for science, and enrolling both his children in the American School in Quito was a sort of natural result of his political and cultural convictions.[17]

In those years Gian-Carlo learned Spanish and English and, greatly stimulated to pursue his philosophical interests, continued his reading of the texts of Benedetto Croce (1866–1952).[18] This youthful interest in Italian Idealism probably influenced his lifelong interest in history, to which his mature writings attest.[19] Moreover, this can help explain why, in his attempt to find a new philosophical perspective on the opposition between idealism and empiricism, he was always far closer to the former than to the latter. In fact, if, on the one hand, we can say that between this Scylla and Charybdis he sought to avoid the one he considered more dangerous, on the other hand, it is clear that the charges of idealism made against Edmund Husserl (1859–1938), one of his principal philosophical points of reference, never worried him in the least.

During his education in Ecuador he distinguished himself in mathematics, in which he was unfailingly first in his class, attaining with relative ease that level of excellence his father demanded. The prize he received in a competition organized for the tenth anniversary of the founding of the Colegio Americano attests to his scholastic merits. On that occasion he submitted a composition that applied psychoanalytical theories to the field of art and literature, showing abilities that an anonymous article published in a Quito periodical judged to be "disconcerting" in a boy of only 18 (Anonymous, 1950). This youthful interest in Freudian theories gave way, in his later years, to a deep distrust of psychoanalysis, with the exception of the currents influenced by phenomenology, such as the one theorized by Jean-Paul Sartre (1905–1980) (Sartre, 1943, pp. 578–595).

After spending four years in Latin America, in September of 1950 Rota went to the United States where he was to remain for the rest of his life, with the exception of periodic trips abroad for work or to visit his family, who in the meantime had returned to Italy. He enrolled in Princeton University at the most fruitful moment of this university's history, when mathematics in the United States reaped the benefits of the presence of those formidable intellectuals who

[16] See Rota Gasperoni (1995), pp. 279–282, and Palombi (1997), p. 267.
[17] Ester Rota Gasperoni, conversation with the author.
[18] Gian-Carlo Rota, conversation with the author.
[19] We also find sporadic references to Croce in his philosophical writings.

had studied and worked in Europe in the first three decades of the 20th century, before emigrating to North America to escape persecutions and the war.[20] At Princeton the senior advisor for his master's thesis (Rota, 1953) was William Feller,[vi] but Rota made the acquaintance of many of the academic personages he described, with great perspicacity and sharp irony, in his celebrated essay *Fine Hall in its Golden Age: Remembrances of Princeton in the Early Fifties* (Rota, 1988a). This and other writings collected in the anthology *Indiscrete Thoughts*, which I edited, along with a number of subsequent interviews, constituted an important part of the documentary material used by Sylvia Nasar to write the book *A Beautiful Mind*, dedicated to the life of Nobel Prize winner John Nash, a friend and colleague of Rota's (see Nasar, 1998, pp. 49–98).

The big names on the Princeton faculty during Rota's student years included some of the greatest mathematicians of the second half of the 20th century: the logician Alonzo Church,[vii] the algebraist Emil Artin,[viii] and the topologist Salomon Lefschetz.[ix] Church in particular had a great influence on the young Rota, not only with respect to mathematics but, more in general, culturally and ethically, teaching him the importance of the relationship between teaching and research. Teaching how to teach is one of the most difficult forms of didactics, and Church's particular type of "metateaching" made a profound impression on Rota. From then on teaching became a fundamental part of his intellectual adventure, to which he devoted himself with great care and discipline. I can say that in the ten years of our working together I never saw him improvise a lesson or a conference. Rather, each one was prepared— theoretically and emotionally—in several hours of solitary concentration, during which he wrote or reread pages of notes in his notebook with the pages marked out in squares, as usually used in elementary school (D'Antona, 1999a, p. 26).

We find a theoretical justification of this habit in Rota's description of Church's manner of teaching, where the great logician's typical lecture is presented as "the literal repetition" of the widely available typewritten text "he had written over a period of twenty years." Rota emphasizes that any doubt about the usefulness of attending such a course would demonstrate a lack of understanding of what really goes on in a university classroom, in that "what one really learns in class is what one does not know at the time one is learning" (Rota, 1988a; see 1997a, p. 5). These lectures are described in an analysis that is so attentive to the most minute details that only Rota's distrust of Freudian theory prevents me from calling it psycho-analytic.[21] During his lectures, Church became "logic incarnate," such that even his nonverbal phenomena (pauses, hesitations, and emphases) were valuable learning

---

[20] I described some aspects of this exodus of the highest representatives of European culture in Palombi (2002), pp. 8–11.

[21] A sort of "clinical" attitude is present also in Rota's description of what may seem to be Church's symptoms of obsession; ibid., p. 4.

opportunities; following his course one learned "to think in unison with him [...] as if following the demonstration of a calisthenics instructor."[22]

In his autobiographical recollections Rota also has high compliments for John Kémeny,[x] one of Church's top former students and colleague, whose seminar in the philosophy of science Rota attended.[23] On that occasion Rota says he learned the necessity of making realistic descriptions of the phenomena to be studied that avoided "black-and-white oversimplifications."[24]

When Rota arrived at Princeton, Albert Einstein (1879–1955) was already quite old. Nevertheless, Rota had a chance to meet him during one of the lectures given by Hermann Weyl.[xi] This extraordinarily interesting series of lessons on the concept of symmetry, involving a vast number of arguments and issues, impressed Rota deeply. Weyl was an important point of reference for the young Rota because, combining scientific with philosophical investigation, Weyl traced that original path of intellectual engagement that Rota himself was to follow. Indeed, we need to bear in mind that Weyl (Weyl, 1952)[25] was not only one of the leading figures of physics and mathematics of the 20th century but he also had a great philosophical temperament, influenced by the ideas of Husserl.[26]

At Princeton Rota continued his reading of Benedetto Croce, thereby laying the groundwork for his subsequent reading of Hegel. He was also able to cultivate his philosophical interests not only by attending specific courses, which, as Rota tells us, at that time did not yet have the great prestige they later acquired, but also by taking part in long discussions with theology students of his own age. Subsequently, Rota decided to minor in philosophy, following the courses of Arthur Szathmary[xii] and John Rawls[xiii]—courses in what can summarily be termed "continental philosophy." Szathmary's teaching distracted Rota from his full immersion in the German Idealist philosophy that had occupied him after his concentration on Croce, turning his attention towards phenomenology.[27] To express his gratitude, Rota later dedicated the anthology of his writings *Indiscrete Thoughts* to Szathmary (Rota, 1997a, p. II).

The reading of José Ortega y Gasset (1883–1955), Maurice Merleau-Ponty (1908–1961), and Sartre fueled Rota's interest in phenomenology, even if the decisive author for his philosophical development continued to be Husserl.[28] Rota recalled that he began reading Husserl's works assiduously in 1957, but did not grasp

---

[22] Ibid., pp. 5–6. Ulam, too, emphasizes the great importance of teaching for a researcher, saying that "the best way to learn a subject is to try to teach systematically. Then one gets the real points, the essentials" (Ulam, 1976; see 1991, pp. 69–81).

[23] Rota (1988a); see (1997a), p. 7.

[24] Ibid.

[25] See also Ulam (1976); see (1991), pp. 69–81.

[26] See Moriconi (1981), pp. 20–22.

[27] Gian-Carlo Rota, conversation with the author.

[28] Gian-Carlo Rota, conversation with the author.

their overall sense until March of 1964. He described the event as a veritable *insight* produced by a reading of the first volume of *Ideas Pertaining to a Pure Phenomenology*;[29] while traveling down the country roads of the Midwest in an old Plymouth driven by his wife, he reread for the umpteenth time a long passage in Husserl's text and finally the sense began to become clear, in an experience similar to anamorphosis.[30]

Martin Heidegger (1889–1976), too, played an important role in Rota's later philosophical studies, but Rota always read him through the interpretative grid of Husserl's work. It is also important to note that he was able to read these philosophers—Husserl, Heidegger, Ortega, and Sartre—in their own languages. Rota, in fact, at the conclusion of his studies, could speak English, Italian, Spanish, French and German, and read Ancient Greek and Latin.[31]

In 1953, after earning his bachelor's degree at Princeton, Rota enrolled at Yale, which had one of the most active schools of functional analysis. There he completed his PhD in 1956, under the guidance of Jacob Schwartz (Rota, 1956). That same year he married Teresa Rondón-Tarchetti; the couple was divorced in 1980. Rota always maintained an attitude of great reserve and discretion with regard to this chapter of his private life. In 1956 and 1957 he worked at New York University as a Postdoctoral Research Fellow with the Courant Institute of Mathematical Sciences, and for the two years after that he worked at Harvard as Benjamin Peirce Instructor. In this period Rota went to the Boston area for the first time. It was love at first sight and immediately became his home, which for the rest of his life he left only for short periods of time. He often liked to remind his friends and students that the Boston metropolitan area has one of the greatest concentrations of universities and research institutes in the world. Between 1959 and 1965 he took his first teaching positions at MIT, made the acquaintance of Norbert Wiener (1894–1964), and struck up a friendship with John Nash (b. 1928). Rota considered Nash to be the most profound mathematician he ever had the good fortune to meet (Rota, 1998b, p. 76), and with whom he frequented not only the staff room of the Mathematics Department but also parties and receptions.[32] Nash's illness and the successive dramatic events, recounted in *A Beautiful Mind*, did not change Rota's opinion, and Rota attempted to keep in touch with him as much as possible even during Nash's first stays in psychiatric hospitals.[33]

---

[29] Husserl (1913). Rota held this text in extremely high esteem. I possess a personal copy of his in the French translation by Paul Ricoeur, with Rota's handwritten notes in the margins. Rota maintained that Ricoeur's translation was an excellent example of successful philosophical rewriting of a Husserlian text.

[30] In particular he mentioned looking at "those multicolor images that conceal a figure, which one sees only if one observes [...] in a certain way" (Rota, 1998e, p. 260).

[31] In Quito he took private German lessons with an elderly Austrian emigrant woman (Ottavio D'Antona, conversation with the author).

[32] Nasar (1998), pp. 239–247.

[33] Ibid., pp. 253–261.

Rota became a US citizen in 1961 and his name, due to a clerical error, acquired a hyphen, taking on the peculiar "Gian-Carlo" spelling that, from force of habit, he ultimately decided to keep.[34] From 1962 to 1965 he was Assistant Professor of mathematics at MIT, and then worked at Rockefeller University until 1967.

Nineteen sixty-four was a particularly important year for Rota, from both a scientific and a personal point of view, marked by the publication of his *Foundations of Combinatorial Theory I* (Rota, 1964), and by his meeting Stanislaw Ulam.[xiv] This work, considered the beginning of combinatorics understood as a coherent theory,[35] made Rota into one of the chief exponents of a branch of research that had long been viewed as a strange and peripheral branch of mathematics. Rota, by contrast, deemed combinatorics to be intimately connected with the main developments of contemporary research and deserving of both individual and academic attention and investment,[36] since "rarely, except perhaps in number theory, has a branch of mathematics been so rich in relevant problems and so poor in general ideas" (Rota, 1995, p. 290). In fact, as he put it, nearly all classical mathematics is applicable to the real world only if this world is constituted by objects and events of a continuous character. But in many situations one can realistically speak only of collections of objects of a discrete character, which act in combinations, one step at a time; the mathematics that is appropriate to such situations is called combinatorial analysis, or combinatorial theory (see Rota, 1969, p. 52).

In 1964 Rota also went to New York for a conference, where he met Ulam. They soon became great friends and associates[37]—a relationship that left a deep mark on the lives of both. Rota wrote that "thinking back and recalling the ideas, insights, analogies, nuances of style that I drew from my association with him for 21 years, I am at a loss to tell where Ulam ends and where I really begin" (Rota, 1987b).[38] This consideration is confirmed in a passage in which Rota examined Ulam's philosophical positions (officially he professed to agree with the positivism of the Vienna Circle), judging them to be "closer to Husserl's phenomenology."[39] Ulam agreed that their association was a sort of intellectual osmosis to the point that, to describe their relationship, he coined the expression "influencer and influencee." In this way Ulam can describe Rota as one of his best influencees and Stefan Banach[xv] as one of his influencers (Ulam, 1976).[40]

---

[34] Gian-Carlo Rota, conversation with the author.

[35] Goldman (1995), p. XVI.

[36] Loc. cit.

[37] See Ulam (1976); see (1991), pp. 72, 104, 129, 207, 263, 268, 328. Ulam expressed his extremely high regard for Rota, judging him to be an excellent mathematician and a friend capable of easing, intellectually and humanly, the void caused by the death of John von Neumann (Ulam, 1976); see (1991), p. 264.

[38] See (1997a), p. 84. See Bednarek, F. Ulam (1990).

[39] Ibid., p. 82.

[40] See (1991), p. 264.

In his essay *The Lost Café*, Rota gives us a portrait of Ulam that is quite different from the official one (Rota, 1987b).[41] If we fully accept Rota's description in *The Barrier of Meaning* of his conversation with Ulam about artificial intelligence, we can only conclude that much of Rota's thinking on this theme comes straight from Ulam.[42] Their discussion of this subject must unquestionably have been an important one, in part because Ulam had worked with John von Neumann,[xvi] one of the fathers of the computer, and thus had been able to follow personally the birth and the principal phases of the development of informatics.[43]

Ulam also played a key role in sponsoring Rota's participation at the Los Alamos Laboratory that began in 1966 (Ulam, 1976).[44] Rota's work at Los Alamos (regularly during August, and when necessary in other periods of the year) continued right up to his death in 1999. His Los Alamos experience also led to his close association with various government services and agencies—in particular the National Security Agency (NSA), which awarded him its Medal for Distinguished Service in 1992 (Rota, 1996, p. 2).[45]

Rota's phenomenological convictions seem to have been further consolidated by an exchange of letters with Kurt Gödel,[xvii] which took place after the publication of an article in which Rota extolled Husserl as "the greatest philosopher in history."[46] In one letter the great logician Gödel speaks highly of Rota's work, sharing his esteem for the founder of phenomenology, but also specifying that, in Gödel's personal classification, Husserl was second to Leibniz.[47] From this personal experience, and through the reading of his works, Rota concluded that phenomenology represented a sort of reference point for Gödel's work even if it was never mentioned explicitly.[48]

In 1967 Rota assumed the editorship of the journal *Advances in Mathematics*, a position he held for the rest of his life, together with his post as editor of *Advances in Applied Mathematics*, which he founded in 1979. Over the years, both journals

---

[41] See (1997a), p. 63 ff. This portrait of Ulam as a sort of phenomenologist manqué is the result of the personal relationship between the two men. In fact, Ulam never uses the term "phenomenology" in his entire autobiography, and the only mention of Husserl is in direct connection with Rota (Ulam, 1976; see 1991, p. 264).

[42] Compare the themes proposed in Rota (1985c) with those in (1985e).

[43] Sokolowski, too, maintains that the "phenomenological" reflection on artificial intelligence can in some way be attributed to the dialogue between Ulam and Rota; in particular, he maintains that this dialogue was extremely important, since the programs used by artificial intelligence are coming up against that which Rota and Ulam termed the "barrier of meaning" (Sokolowski, 1992). See also Sokolowski (1988).

[44] See Ulam (1991), p. 264.

[45] See also Cerasoli (1999), p. 19.

[46] Rota (1998e), p. 261. Unfortunately I have not been able to track down the essay to which Rota refers.

[47] Loc. cit.

[48] Loc. cit.

came to play leading roles on the scene of mathematical publications. These two periodicals were also of importance for my own specific interests, because it was in their reviews of mathematics and philosophy books that Rota shot his poisoned arrows against analytic philosophy.

In 1967 Rota also definitively returned to MIT, creating around him a group of students and colleagues who represented what was described as "the Cambridge School of Combinatorics."[49] Goldman remembers the years up to 1971 as an exhilarating period, during which the MIT weekly seminar was the center of a vast field of combinatory research carried on in the Boston area.[50]

From 1972 on, Rota taught at MIT in the dual role of mathematician and philosopher. One of the reasons he chose to take a permanent position at MIT was that, unlike other major universities, MIT permitted him to teach philosophy, and not mathematics only.[51] At MIT his research and teaching in phenomenology was completely independent of the philosophy department, which was marked by analytic philosophy. The only philosopher at MIT with whom he had a relationship of mutual esteem was Thomas Samuel Kuhn (1922–1996), who taught at MIT for many years as Professor of Philosophy and History of Science.[52] Rota also enjoyed a personal and professional relationship with Robert Sokolowski (b. 1934), an important Husserl scholar who taught at the Catholic University of America in Washington, DC.[53]

In 1975 Rota delivered a series of lectures organized by the National Academy of the Lynxes,[54] one of the oldest and most prestigious Italian cultural institutions, and took part in the congress of the Society for Phenomenology and Existential Philosophy held in Nashville, an experience he repeated three years later in Pittsburgh. In August of 1989 he was a speaker at the annual meeting of the Husserl Circle held at Fort Collins, Colorado,[55] and in December of that year he was invited to give a lecture, titled *Les ambiguïtés de la pensée mathématique*, at the Collège de France (Rota, 1990a). He also participated in major conferences on November 17, 1992, at the Boston Colloquium for the Philosophy of Science (Rota, 1997d), and on March 2, 1995, in Memphis at the Eighth International Kant Congress (Rota, 1997g).

Rota's growing international renown induced Italian universities and institutions to invite him to Italy more and more frequently, first as a mathematician,

---

[49] Goldman (1995), p. XVI.

[50] Loc. cit.

[51] Gian-Carlo Rota, conversation with the author.

[52] Gian-Carlo Rota, conversation with the author.

[53] See Rota (1988b), pp. 376–386. Rota participated in the Congress in honor of Sokolowski held on November 12, 1994, at Catholic University, with a paper published in Rota (1997h). Sokolowski, moreover, dedicated his book *Pictures, Quotations, and Distinctions* (1992) to Rota. Rota mentioned him for the first time in Rota (1973a).

[54] Rota (1997n).

[55] Rota (1989b).

and later also as a philosopher. It seems that the first public initiative of a philosophical nature in which Rota participated in Italy was a congress held in 1976 in Perugia at the convent of Monte Sant'Angelo (Cerasoli, 1999, p. 16). He later held numerous series of lessons and conferences in the degree and doctoral programs in mathematics, informatics, and philosophy.

From 1982 he was a member of the National Academy of Sciences, where he was president of the mathematics section from 1994 to 1997, and in 1986 he was Lincean Professor at the Scuola Normale Superiore in Pisa. The extraordinary variety of Rota's interests and international relations is attested to by the honorary degrees he was awarded by European universities for mathematics[56] and for the Sciences of Information.[57] In Italy he was a member of the scientific committee of the Scuola Normale in Pisa for several years, and of that of the Libero Istituto Universitario Carlo Cattaneo, where he also delivered the inaugural lecture in 1991 (Rota, 1992b). In December of 1990 he began his association with the Istituto Italiano per gli Studi Filosofici in Naples (his first series of lectures was titled *Phenomenology and Foundations of Mathematics*), where he continued to lecture nearly every year until 1998. The most important of these lessons were later collected in the anthology *Lezioni napoletane*, which the Istituto published after his death (Rota, 1999a). Rota also delivered a lecture, *Phenomenology, Mathematics and Culture in the Sciences*, in Milan in 1993 at the Università Statale, on the occasion of the publication of his first anthology of philosophical writing, titled *Pensieri discreti* (Rota, 1993a).

For many years Rota returned to Italy for his summer and winter vacations to visit his relatives, especially his mother until her death in 1991. In these periods he cultivated his numerous interests and, in particular, followed the group of mathematicians who were his students, of whom he was especially proud. This group arose as a result of a series of lectures that Rota delivered at Cortona, in Tuscany, beginning in 1974, in the summer school program sponsored by the C.N.R. (the Italian national research council) (Senato, 1999). Similar groups arose in other countries as well, where his teaching encouraged the development of combinatorial studies.

Rota's human and intellectual journey came to a sudden end on April 18, 1999, in his home-cum-library in Cambridge, where a heart attack made him cease to calculate, philosophize, and live. His ashes are buried in Mount Auburn Cemetery, near the Charles River, in the quiet of Cambridge.

1.2 ✧ A Philosophical Style

Rota's mathematical and philosophical investigation unfailingly influenced one another reciprocally. This accounts for the characteristic style of his texts, which

---

[56] The Universities of Strasburg (1984) and of L'Aquila (1990).

[57] The University of Bologna (1996).

focus on the rewriting of the great texts of phenomenology with very little regard for quotations, bibliographical references, and historical contextualizations.[58] Moreover, this style is motivated by the eminently theoretical character of his philosophical activity, stemming from problems of mathematical research and devoid of specific historico-philosophical interests.

For Rota, the names of authors are indexes, and the texts are toolboxes, tools,[59] and, simultaneously, the material to dismantle and "recycle" in order to criticize what he considers to be the "vices" of contemporary empiricism. His caricatural description of the scientistic philosophies, whose defects he magnifies to make them more easily comprehensible, is instrumental to his criticism of the "ingenuous dogmatism of objectivism."[60] This particular philosophical style can be considered as the product of his activity as a mathematician, of his distinctive reading of Husserl's and Heidegger's texts, and of the cultural context in which he lived and worked.

To understand the influence of certain aspects of mathematical practice on Rota's phenomenological investigation we must begin with what he considered to be the essential discrepancy between the ways research is presented in philosophy and in mathematics. Mathematicians do not teach mathematical analysis to their students by discussing the original texts of Newton, Leibniz, and Torricelli; neither do they consider the first works written on the subject to be classics of a theory.[61] In fact, while the proofs of the discoverers are often clumsy and redundant, a text considered classic is the successful completion of a series of progressive steps in the understanding of a mathematical theory (Rota, 1990a).[62] Every mathematician knows that a new exposition of an already discovered theory represents a work of very high-level research; it is not fortuitous that most of the articles published in the journals of mathematical research consist precisely of such expositions.[63] But in the case of philosophy, for Rota, the situation is completely different.[64] If philosophers intend to teach their students phenomenology, they have no choice but to invite them to read the original writings of Husserl. In fact, Rota is convinced that the countless exegeses—the philological and historical analyses of Husserl's

---

[58] *Indiscrete Thoughts* (1997a) is an exception; after a heated discussion with Rota, I managed to convince him to include a few bibliographical references.

[59] I shall consider the importance of the concept of tool in Rota's thinking in Section 2.4, which is dedicated to the analysis of the relation of foundation.

[60] This definition is the title of Lesson XII of Husserl (1956).

[61] Gian-Carlo Rota, conversation with the author.

[62] See Rota (1997a), p. 109.

[63] Ibid., pp. 113–117. This theme will be discussed in depth in Chapter 3.

[64] Rota (1990b); see (1997a), pp. 93–94. From an epistemological standpoint Kuhn provides an explanation of this situation; in fact "in history, philosophy and the social sciences [...] the elementary college course employs parallel readings in original sources, some of them the 'classics' of the field" (Kuhn, 1962, p. 165).

writings—even if of very high value, practically never represent full and proper treatises of phenomenology that intend to teach the subject matter in the same way in which one teaches mathematical analysis to students. This is why he endeavors to rewrite Husserl's texts with the spirit of a mathematician who strives to grasp the sense of a mathematical theory.[65]

Later on I shall examine in depth the motivations and the consequences of this theoretical position.[66] For now it is important to emphasize that for Rota, the meaning of a mathematical theorem cannot be individuated in the final purpose or in the biography of its discoverer. To understand a theorem, such aspects are negligible or at most subsidiary, even if by this Rota does not intend to underestimate the historical dimension of knowledge. On the contrary, he interprets this dimension in terms of ideal and not factual reconstruction. For him, the theme of rewriting has a more general value, which also involves the translations of his works into languages he knew well, such as French, Spanish and, obviously, Italian. It may seem strange, but, despite his thorough knowledge of the Italian language, Rota always refused to write in Italian, theorizing the impossibility of writing at a philosophical or an artistic level in two languages. I can personally attest to the difficulty of translating his writings, due not only to obstacles of a conceptual nature but, above all, to his rigor as a writer and a philosopher, which drove him to go over the translation personally on what he considered a fresh occasion for investigation and thought. This is why the Italian translations of his works are often so different from the English original that they turn into independent works. In many cases the translator, explicitly encouraged by Rota, added considerations, personal reflections, and quotes, and got so involved in the rewriting of the text as to effectively become a coauthor.[67] Rota, in some way, appears to express a demand that is not very far from Benjamin's principle that a work of translation must not shirk the duty of perfecting a text (Benjamin, 1955).

Phenomenology also influenced Rota's writings in a sense as particular as it was profound. There are certain philosophical currents that reduce scientific enterprise to pure technique, basing themselves upon suggestive (albeit often inappropriate) quotations from the phenomenological—especially Heideggerian—literature, with the purpose of demonstrating that science "does not think," or at least that there is a basic incompatibility between science and philosophy. By contrast, Rota's investigation is based on a diametrically opposed conviction; he is persuaded of the depth that sustains any genuine scientific discovery, and of the possibility of coupling

---

[65] Rota suggested that Sokolowski write a text of introduction to phenomenology for this very purpose; see Sokolowski (2000), pp. IX, 1.

[66] See Chapter 3.

[67] Wherever possible, I have attempted to take this into account by indicating the various versions of an original text in the bibliography.

epistemological[68] research with "uncomfortable" philosophical traditions such as the phenomenological-hermeneutic.

From this point of view Rota's position is close to that of Ricoeur's, which interprets hermeneutics as a reflection on the conditions of philosophical possibilities of epistemology.[69] Of these conditions the most important is represented by "sense," understood as the primordial dimension on the basis of which it is possible to articulate the understanding of every cultural and natural phenomenon. The primordial belonging to a world, to a horizon of sense, is the condition of possibility of every theory of knowledge.

Pursuing this perspective, Rota uses Husserl's texts as a point of reference and interpretative grid through which to filter texts and reflections of other philosophers, beginning with Gilbert Ryle (1900–1976) and Ludwig Wittgenstein (1889–1951).[70] Both these authors present important points of contact with phenomenology. Ryle in particular followed its developments, presenting a text (Ryle, 1932)[71] in the United Kingdom that influenced the reception of Husserl's thought in the English-speaking world.[72] The Husserlian influence accounts for the character of "work in progress" of Rota's writings, founded on the method of phenomenological description and thus always incomplete and "open-ended by their very nature."[73]

Rota maintains that the value of phenomenology does not reside in illusory definitive conclusions but is founded on the discourse and on the sense of this very philosophical perspective. Phenomenological descriptions are open-ended because in the future it will always be possible to uncover characteristics that have been neglected not necessarily due to a lack of rigor. It is a question, rather, of a constitutive incompleteness; a *neglected* characteristic is such only *a posteriori*, because we can (and must) undertake further phenomenological analyses of it only by virtue of "renewed needs that we will have which we cannot foresee now."[74]

---

[68] I shall use the term *epistemologia* in a broad sense. Clearing the field of its etymological meaning (which refers to an incontrovertible knowing), Francesca Bonicalzi indicated that it "modulates on the corresponding French and English terms *épistémologie* and *epistemology* that translate that *Wissenschaftlehre* which [...] characterizes German philosophy from Fichte to Husserl." Contemporary philosophy uses the term in two ways: the first, in a positivist sense, that "suggests [...] questions that regard the unity of knowledge [...] in its relation to the plurality of sciences"; and a second, of Kantian origin, that "brings into play the properly epistemological question of the subject of knowing and of the objectivity of knowledge" (Bonicalzi, 1997, p. 15). It is above all the second sense that inspires Rota's investigation.

[69] See Ricoeur (1981).

[70] Rota makes reference to that current of interpretation, which relates the second Wittgenstein to phenomenology; see Monk (1990), pp. 489–519.

[71] Ryle (1962) represents another important essay on the subject.

[72] See Costa (2000f), p. 301.

[73] Rota (1991a), p. 75.

[74] Loc. cit.

I have already indicated that Heidegger's reflection interests Rota in so far as it represents a development of Husserl's own thinking. In particular, Rota maintains that the "ontic-ontological methodology of *Being and Time* is derived from Husserl's descriptions, but Heidegger uses the Husserlian discovery of new relations of being in a different and more radical way."[75] This somewhat "evolutive" reading was in fact suggested by Heidegger himself who, in his last writings, explicitly recalls his debt to his old master.[76] Nevertheless, this interpretation, which affirms the continuity of phenomenological investigation, is not the only lens through which to read the relationship between the two great German philosophers. In fact, apart from the question of the personal relations between them, we must recall that the theoretical tension grew so great that they found it impossible to agree on a definition of the very term "phenomenology."[77] Rota's evolutive reading makes him one of the few scientists who dialogues with Heideggerian reflection, "purifying" it of all antiscientific positions and valorizing often neglected aspects of Heidegger's development. In this regard we must recall that for Rota, the first phase of Heidegger's thinking

> is contemporary with the forward leap in logic that began with the work of Frege and the symbolic notation invented by Peano at the turn of the century. Young Heidegger's book review shared the enthusiasm for the new logic. (Rota, 1987a, p. 122; see 1997a, p. 188)

This thesis is sustained by a short essay (Heidegger, 1912) in which Heidegger shows he is well acquainted with the fundamental conquests of logic at the beginning of the 20th century, quoting works by Bernhard Bolzano (1781–1848), Gottlob Frege (1848–1925), Henri Poincaré (1854–1912), and Bertrand Russell.[78] These symptoms of his interest in contemporary science are also present in his principal work, *Being and Time*, in which he demonstrates his close attention to the crisis of the foundations of science (and of mathematics in particular).[79] I also wish to recall that for Rota,[80] the so-called "turning" does not completely modify certain

---

[75] Rota (1987a), p. 124; see (1997a), pp. 188–91. The latter, *Three Senses of "A is B" in Heidegger*, is a considerably abridged version of an earlier text in Italian. Quotations in this book are translated from the Italian text when they are not found in the English version.

[76] See Heidegger (1969).

[77] See Heidegger and Husserl (1962).

[78] See Marini (1982b), pp. XIV–XVII.

[79] See Heidegger (1927), pp. 29–30.

[80] "I would like to advance a thesis that may seem reactionary [...] according to which Heideggerian thought after the *Kehre* [turning] in no way modifies the ontological law of the 'as' [...] which holds firm up to the very end of Heidegger's philosophical development" (Rota, 1987a; see 1997a, pp. 188–91). The *Kehre* is "Heidegger's turning away from the analytics of Dasein toward the question of Being" (Gasché, 1986, p. 85). Heidegger realized after 1927 that the legacy of metaphysical language deeply conditioned his philosophical research, so he began to write using a new style that showed his turning in thought.

fundamental aspects of Heidegger's philosophy; in this case the evolutive thesis is advanced within Heidegger's reflection to safeguard a certain link with his youthful interest in science.

Rota's reading presents a Heidegger in some sense "domesticated" and "refined," designed for epistemological use, but no less interesting on this account. In his effort to update the Husserlian philosophy, Rota utilizes the reflections and the texts of Husserl and Heidegger as complementary instruments in dealing with the great problems of phenomenology in the contemporary cultural and scientific context.

We find another Husserlian influence in an important "terminological" precaution, when Rota indicates that the "spirit" of phenomenological investigation can easily be misunderstood. In particular he fears confusion between phenomenological evidence and the vulgar understanding determined by common sense drenched in objectivistic vocabulary. Rota deals with this problem in a sophisticated manner, recalling the Heideggerian lesson that shows the impossibility of completely and definitively overcoming metaphysical language, of which objectivism, too, is an expression. Heidegger wrote that *Being and Time* remained incomplete precisely because of the inadequacy of the language that contemporary philosophy inherited from the metaphysical tradition (Heidegger, 1967, pp. 239–276). It is well known that the etymological meaning of the term "metaphysics" is rendered with the expression "beyond physics." In the prefix "meta" it is possible to recognize a strategy of overcoming and going beyond, which, as the case may be, can be referred to nature, to sensible appearances, or to particular sciences. This strategy individuates a common denominator of the entire Western philosophical tradition with respect to which phenomenology takes a step backward. In this sense the project of overcoming metaphysics comes within the selfsame metaphysical tradition that one wishes to relinquish, and cannot be achieved because it is intrinsically contradictory.

Rota attempted to get around this obstacle, to *force* this limit, by insisting on the constitutively provisional and precarious nature of any reform of the philosophical vocabulary and, in particular, of the one he himself had undertaken. In this perspective the transformation of objectivistic vocabulary has to be conceived of as an inexhaustible demand and not as a definitive result. Husserl himself had this to say about the instability of phenomenological language:

> Because it will not do to choose technical expressions that fall entirely outside the frame of historically given philosophical language and, above all, because fundamental philosophical concepts are not to be defined by means of firm concepts identifiable at all times on the basis of immediately accessible intuitions; because, rather, in general long investigations must precede their definitive clarifications and determinations: combined ways of speaking are therefore frequently indispensable which arrange together a *plurality* of expressions of common discourse which are in use in approximately the same sense and which give terminological pre-eminence to single expressions of this sort. (Husserl, 1913, p. XXIII)

Rota's concerns are quite similar to Husserl's when he seeks to distance himself from the linguistic misunderstandings generated by common sense, and in particular by that which he describes as the empiricist tradition. For these reasons he holds that

> the best we can do is to display precautionary signals within our discourse, which means that we are forced to use words of clearly objectivistic origin. Words like "problems," "solutions," "arguments," and "relationship" which are used repeatedly in our philosophical discourse must not be interpreted in their conventional sense when they are broached in phenomenological language. (Rota, 1987a; see 1997a, p. 188)

The same precautions have to be utilized for the terms "relation," "correspondence," and "function" that we find within his writings. Nevertheless, these precautions do not prevent numerous variations and shifts of meaning from entering Rota's vocabulary. This is due not only to the attempt to "force" the metaphysical and objectivist tradition described above, and to the temporal distance between the earlier and later writings, but also to sudden second thoughts and attempts to criticize his own thinking. Rota did not like flatterers and spurred himself and his coworkers on (always courteously, often insistently) to criticize, and to bring to light the defects of his findings.[81]

To understand Rota's texts better one must also bear his teaching activity in mind, and in particular the course of Phenomenology at MIT that played a fundamental role in his philosophical work for over 20 years. The public that attended these lectures received an education that was profoundly different from European historical and philosophical culture.[82] This particular environment had a marked influence on his style; as a scientist teaching a class of science-oriented students, in order to be understood he had to draw his phenomenological descriptions and examples from scientific literature, and to present them in the closest way possible to the manner of teaching scientific subject matter.

This particular situation (along with the influence of Husserl's writings) accounts for one of the most characteristic aspects of Rota's style—namely, the importance of examples. His argumentations are not specifically characterized by broad generalizations but rather by the cadenced pace of exemplary cases that he preferred to the "myth of definition" (Rota, 1991a, pp. 20–23; Rota, 1990b, pp. 97–98). While common sense believes that everything can be defined—that "their essence can be captured in words,"[83]—Rota insists that "learning concepts is a complex feedback process, and defining the concept is only one of the steps."[84]

---

[81] Palombi (1997) is the fruit of a process of this sort.

[82] I myself experienced this cultural and human context personally when I followed Rota's lectures on Phenomenology in the autumn of 1990.

[83] Rota (1991a), p. 21.

[84] Ibid., p. 22.

This does not mean that definitions are useless, but rather that their importance is reappraised and reduced to *one* of the possible points of departure (or of arrival) of the various experiential processes through which comprehension is achieved. Let me make clear that Rota, in speaking of "complex feedback," followed a specifically phenomenological perspective;[85] this specification provides us with the fundamental coordinates to interpret the circularity that characterizes the feedback of experience, without reading this expression equivocally in cognitivist or, more generally, analytic terms.

Hence, in the phenomenological horizon, definitions have to relinquish all privilege to the advantage of examples, in the conviction that genuine understanding stems only from repeating certain exemplary cases and extrapolating the characteristics they have in common. Rota's style characterized by examples is thus to be seen against the background of a principle we could define as "methodological," with consequences we can reduce to two orders. The first is represented by the didactic potentiality of his writing and his argumentations, which render authors of extraordinary importance and difficulty such as Husserl and Heidegger accessible even to neophytes and beginners.

There is, however, also a second order of consequences, connected with the relationship of his texts with scholars accustomed to a different sort of scientific exposition based on quotations, bibliographies, and notes. In the writing of this book I have had to cope with the difficulties of texts often devoid of these elements in my endeavor to reconstruct, wherever possible, the sources, references, and allusions to the philosophical texts Rota used. This effort made it possible for me in some cases to develop and broaden his philosophical reflections by bringing out themes that are only implicit in his text.[86]

1.3 ✧ Rota and Analytic Philosophy

The general picture of Rota's philosophical style is not complete unless we consider the fact that the cultural context in which he lived and worked was characterized by the opposition between analytic and continental philosophy. He was particularly influenced by this clash, whose correlates are historical, thematic, argumentative, and even include writing style.[87] Rota sided with the continental "party" and took a position that was highly critical of analytic philosophy (in all four of its aforementioned aspects) in writings that provoked heated polemics,

---

[85] Here, feedback has to be interpreted in the light of Husserlian *Rückfrage*; see Derrida (1962), p. 50.

[86] A concise bibliography included in the Program of the two courses Rota gave every other year at MIT (24.171 *Introduction to Phenomenology* and 24.172 *Being and Time*) contained in Rota (1994b), p. 522, presents the following texts: Heidegger (1927), Husserl (1900–1901), (1913), (1950a), and (1950b).

[87] See D'Agostini (1997), pp. 57–58.

but above all, in his teaching, his lectures, and in his discussions with friends and colleagues. To present his philosophical perspective on this subject I shall thus draw also on accounts and recollections of persons who, like me, witnessed his polemics firsthand.

Although analytic philosophy unquestionably owes a great deal to logical positivism, the two currents also possess numerous and pronounced distinctions. Indeed, these historiographical labels conceal a complex movement that cannot easily be reduced to stereotypes.[88] What is more, logical positivism itself went through an extraordinary evolution known as its process of "liberalization," which led the exponents of this current very far from their original positions, as the intellectual adventure of Rudolf Carnap (1891–1970) demonstrates.[89] Rota, by contrast, was not particularly attentive to these important historical and theoretical distinctions, and tended to identify logical positivism with analytic philosophy, or at most to present the latter as heir and continuer of the former.

The force of his polemic against the analytic current is attested to by an episode known to only a few people. Rota's contribution to the historical catalogue published by *il Saggiatore*, which I translated, originally began by criticizing the "dark years of Carnap and Reichenbach," and only in the final proofs was transformed into a more muted reference to the "dark years of reductionism" (Rota, 1998a, p. 93).[90] There is a trace of Rota's second thoughts in the Introduction by the editors who, unaware of this last-minute correction, take note of his polemic against "positivism and its 'new adepts'" (Cadioli et al. (Eds.), 1998b, p. 14). This correction shows, moreover, that Rota tended to trace analytic philosophy back to (if not to identify it with) another favorite butt of his polemics—reductionism. In this way he wanted to highlight that which is considered one of the chief characteristics of analytic thought—its concentration on particular aspects of reality, on parts torn from their comprehensive contextual relations.[91] Furthermore, probably taking very seriously (perhaps almost literally) the philosophical genealogy contained in *The Scientific Conception of the World*,[92] Rota traces a descent that goes from the British empiricists to those whom they consider their contemporary logical positivist epigones.

If logical positivism is a convenient target, it is also true that, more easily than other currents, it has lent itself to the role of an unconscious metaphysics for scientists and popularizers who are often far removed from philosophical

---

[88] Ibid., p. 62.

[89] I took up this question from the particular perspective of the epistemological debate on the scientificity of psychoanalysis; see Palombi (2002), pp. 40–43, 70–72, 83–84, 93–94.

[90] Obviously Rota was referring to the epistemologist Hans Reichenbach (1891–1953), leading spirit of the Berlin Circle and editor, along with Carnap, of *Erkenntnis*, one of the most important journals of the logical positivist movement. See also Mugnai (2009), pp. 241–243.

[91] See D'Agostini (1997), p. 60.

[92] See Carnap, Hahn, Neurath (1929).

investigation.[93] For this reason Rota criticized some logical positivist positions as scientistic particularly in reference to the ideological distortions they produced among scientists. An example of this attitude is represented by Ulam who, as we saw, Rota criticized for his philopositivist declarations.

Rota believed he had found a fundamental reference for the reconstruction of the analytic current's philosophical genealogy in the figure of Wittgenstein, whose *Tractatus Logico-Philosophicus* he takes into particular consideration (Wittgenstein, 1922). In his view this text is at the root of "one of the most insidious prejudices of the twentieth century" (Rota, 1990b),[94] referring in particular to the celebrated thesis that "what can be said at all can be said clearly; and whereof one cannot speak thereof one must be silent" (Wittgenstein, 1922, p. 27). Rota interprets drastically Wittgenstein's entire opus after the *Tractatus* as the attempt to amend this thesis, and in particular he rereads *Philosophical Investigations* (Wittgenstein, 1953) as "a loud and repeated retraction of his earlier gaffe" (Rota, 1990b).[95]

Rota takes as a model of nonreductionist philosophical investigation (and thus close to phenomenology and to its descriptions) a metaphor extrapolated precisely from *Philosophical Investigations*. He compares the concepts of philosophy to "the winding streets of an old city" that, to be appreciated, demand a comprehensive vision and a patient investigation. To understand their topography we must "stroll through them while admiring their historical heritage."[96] This metaphor represents the synthesis of that which he sees as essential parts of phenomenology—attention to detail and nuance, never exhaustive, and always susceptible to new investigation. One of the few exceptions to this attitude is represented clearly by his drastic anti-analytic positions that, at times, tainted him with the same vices for which he reproached his adversaries.

It is legitimate to wonder about the reason for this attitude, which cannot be attributed exclusively to his continental education or his personal idiosyncrasies. His personal statements and some of his writings point to a climate of intellectual and academic dispute that seems to lend credence to Richard Rorty's (1931–2007) conviction that the analytic philosophers exercise a sort of hegemony in the philosophy departments of universities in the United States, to the point of "exiling" continental philosophers to the literature departments.[97] For this reason, and not fortuitously, Rota often launched resounding Philippics against "the Anglo-Saxon

---

[93] In this regard I refer the reader to the analysis of some of Eysenck's philosophical theses in my Palombi (2002), pp. 35–38.

[94] See Rota (1997a), p. 95.

[95] Loc. cit.

[96] Loc. cit. Wittgenstein writes that "our language can be seen as an ancient city: a maze of little streets and squares, of old and new houses, and of houses with additions from various periods; and this surrounded by a multitude of new boroughs with straight regular streets and uniform houses" (Wittgenstein, 1953, pp. 8,18).

[97] See Rorty (1982).

philosophical orthodoxy" that imposes "severe penalties (the loss of one's chair, exile, exclusion from the best 'clubs') on those who do not conform" (Rota, 1998a, p. 94). He maintains that in the Anglo-Saxon countries

> the classical problems of philosophy have become forbidden topics in many philosophy departments. The mere mention of one such problem by a graduate student or by a junior colleague will result in raised eyebrows followed by severe penalties. In this dictatorial regime we have witnessed the shrinking of philosophical activity to an impoverished *problématique*, mainly dealing with language. (Rota, 1990b; see 1997a, p. 98)

I myself witnessed a polite but decidedly heated exchange of opinions between Rota and Willard Quine (1908–2001) during a reception prior to a conference held at the American Academy in Boston in the autumn of 1990. The two philosophers taunted one another for quite some time, and at the conclusion of the conference Quine theatrically invited me to follow only Rota's mathematical, and not his philosophical, research. Perhaps the elderly American philosopher had not forgotten that Rota, some years before, in reviewing one of his books (Quine, 1966), remarked that "when a philosopher writes well, one can forgive him anything, even being an analytic philosopher" (Rota, 1978c; see 1997a, p. 255).

Nevertheless it was not the academic tensions and personal conflicts that worried Rota, but rather a cultural transformation that had been produced by the analytic hegemony—namely, the liquidation of the "historicity of thought" operated by distinguished scholars occupying prestigious chairs of philosophy in the United States. This operation led to the modification of university course programs, replacing courses in the history of philosophy, in Greek, and in German with "required courses in mathematical logic" (Rota, 1990b; see 1997a, p. 100).

This statement ought to be backed up by statistical analyses and a comparison of the current course programs in US universities with those of the past, a task that goes beyond our present purposes. Nevertheless, it gives food for thought that of the 56 courses in the domain of philosophy offered by MIT in the academic year 1994–1995 only three of them referred explicitly to the history of philosophy.[98] While it is true that MIT has a decidedly technical-scientific orientation, I think that this fact is significant, and that it confirms Paolo Rossi's reflections on the vicissitudes of the Committee for Pluralism in Philosophy that was instituted in the United States to curb the hegemony of analytic philosophy.[99]

---

[98] The courses were the following: 24.07 *Classics in the History of Philosophy*, 24.200 *Ancient Philosophy*, and 24.202 *Modern Philosophy: Descartes to Kant*. See Rota (1994b).

[99] See Rossi (1995), Vol. I, p. 10.

✧ ✧ ✧ ✧

CHAPTER TWO

A Phenomenological Perspective

"That which had previously been formulated as the problem of the sense of Being has now been reformulated as the pure and simple problem of sense, [...] the word 'Being' has quietly been set aside. Now the question is: where does sense arise from?"

—Gian-Carlo Rota

Rota's philosophical reflection follows the path cleared by Husserl and, in particular, crisscrosses with the problems that emerge in *The Crisis of European Sciences* (Husserl, 1959); namely, that since philosophy originates in a reflection on the problem of sense (Rota, 1991a, pp. 52–56 and 1993a, p. 129),[1] the current crisis of contemporary culture, at once philosophical and scientific, arises precisely from the forgetting of this great problem.

Rota intends to respond to the challenge of a science that "excludes in principle" the questions of sense (Husserl, 1959, p. 6)[2] by shouldering, himself, one of the tasks of Husserlian phenomenology; i.e., by developing a philosophy that wants to be "scientific" without accepting "an objectified [...] and mechanical vision of science" (Franzini, 1991, p. 8). Rota devoted most of his fundamental opus, *The End of Objectivity* (Rota, 1991a) precisely to a reflection on the problem of sense and of its forgetting; for this reason it behooves us to begin this chapter with a preliminary reflection on the title of this book which is, in many respects, programmatic.

2.1 ✧ The Problem of "Sense" and the "End of Objectivity"

The English term "objectivity" possesses the dual meaning of impartiality and of having the reality of an "object" not dependent on the mind. Rota uses the term in both senses to refer to that ensemble of philosophic doctrines that Husserl brings together under the term "objectivism" (see Husserl, 1956, and 1959, pp. 68–70).

---

[1] Rota chooses to use the word "sense" rather than "meaning," saying "we will not use the word 'meaning' because it is tainted with special uses" (1991a, p. 51).

[2] Husserl is concerned not only with the sciences of nature but also with those of the spirit that suffer the same destiny, characterized by objectivism (ibid., pp. 6–7).

It is difficult to understand fully the meaning of this term if we follow Rota's text alone which, poor in definitions by methodological premise,[3] limits itself to defining objectivism (or objectivity) as "the understanding of our world as made up of 'things'" (Rota, 1991a, p. 124). This indication further complicates the situation since the word "thing" (*cosa*) is seen in common language as the most general of all terms, and thus understanding "what [thing] is a thing" (*che cosa sia una cosa*) throws us immediately in a sort of logical circle expressed also by a linguistic contortion. In this case, to clarify Rota's definition I shall not follow his examples in *The End of Objectivity* that, to "win over" the MIT public, lead toward a sort of identification of objectivism with materialism and its mechanistic variant;[4] but neither shall I follow the overly facile Heideggerian assonances. Rather, I believe it is important to return to the texts of Husserl himself.

Husserl uses the term "thing" with different meanings; thus we have to select which is the one that Rota considered to be the constitutive element of the objectivistic world. We need to recall here that Rota used the term "thing" in the meaning given by common sense imbued with scientism, maintaining that the physical and factual world "seemed to be the ultimate—completely *independent* and irreducible, beyond which there is no explanation" (Rota, 1991a, p. 2; emphasis added). This "thing" is characterized above all by its *independence*, and to better understand its meaning we must make reference to that which Husserl defines as the "physicalistic thing."[5]

Objectivism is a philosophical strategy that seeks to account for the whole of reality and of experience through a sort of rational reconstruction that sets out from "physicalistic nature," from the "*thing itself in itself.*"[6] But let us not forget that the things of objective science "are not [...] things [...] like stones, houses or trees. [...] they are 'representations-in-themselves' [...] ideal unities of signification, whose logical ideality is determined by their *telos*, 'truth in itself'" (Husserl, 1959, p. 130). As Husserl further explained, "What characterizes objectivism is that it moves upon the ground of the world which is pregiven, taken for granted through experience, seeks the 'objective truth' [...] what it is in itself" (Husserl, 1959, p. 68). Albeit from another cultural context, the words of Thomas Nagel (b. 1937) help to make it clear (in extreme synthesis) that physical things have been made possible by a

> method of investigating the observable physical world not with respect to the way it appears to our senses [...] but rather as an objective realm existing independently of our minds. [...] In order to do this, it was necessary to find ways of detecting and measuring and describing features of the physical world which were not inextricably tied to the ways things looked, sounded and felt to us. [...] The whole

---

[3] In this regard see Section 1.2.
[4] We recall the classical criticism of mechanism in Mach (1883).
[5] Husserl (1952a), p. 89.
[6] Loc. cit.; emphasis in original.

idea of objective physical reality depends on excluding the subjective appearances from the external world and consigning them to the mind instead. (Nagel, 1994, pp. 65–66)

In essence, we can say that the scientific object is determined as the isolation of a particular layer of reality. This operation determines the epistemological priority of the quantitative dimension to which the qualitative dimension is also reduced. A clock, a stone, or the light of a sunset are not scientific objects as such, but have to become so through the elimination of all the subjective components that "color" ("taint") them.

A great deal has been said about the genesis of this objectivist strategy, with emphasis on the role played by the reflections of Galileo and Descartes.[7] These philosophers acted within the common framework of a project of scientific creation, but there are nonetheless fundamental differences between them. For example, Francesca Bonicalzi emphasized one distinction based on the examination of the expression "removal of the sentient animal."[8] In Galileo this is part of an epistemological strategy that founds scientific discourse on the neutralization of the sense organs in order to "make a distinction between constitutive aspects of the structure of bodies—shape, size, spatial and temporal determination, motion, resistance—and their mechanical action on the organs of sense—tastes, odors, colors, sounds—that have existence only in the body of the animated subject."[9]

The case is significantly different in Descartes, for whom there is "no cognitive experience of sensible reality within which to make a distinction between what is proper to the object and what occurs in the subject. [...] Indeed [...] the expression 'sensible reality' is ambiguous because it leads to the misunderstanding that the body is defined in relation to our senses and not by that which it is; i.e., extended substance."[10] This important difference between Galileo and Descartes articulates and sustains one of the fundamental strategies of objectivist thought, namely, the radical attempt to separate the world from the human being that perceives it. In this perspective the physical thing does not belong to the primordial experience that the human being has of the world, but is produced as the result of a long tradition that posits it "as an empty identical something as a correlate of the identification possible according to experiential-logical rules and grounded through them" (Husserl, 1952a, p. 93).

This reversal that posits physical things as primordial rather than the things experienced in concrete human life, which conceives of them as "in themselves," is at the origin of objectivism and of the consequent crisis of science understood "as the loss of its meaning for life" (Husserl, 1959, p. 5). It is thus clear for Rota

[7] In this regard see Husserl (1959).
[8] Bonicalzi (1998), p. 128.
[9] Loc. cit.
[10] Ibid., pp. 131–132.

that the end of objectivity is the end of the conception of the object as thing in itself, external to and totally independent of a knowing subject. Compare Husserl's view that

> [a]n object existing in itself is never one with which consciousness or the Ego pertaining to consciousness has nothing to do. The physical thing is a thing belonging to the surrounding world even if it be an unseen physical thing, even if it be a really possible, unexperienced but experienceable [...] physical thing. (Husserl, 1913, p. 106)[11]

Phenomenological reflection ought to represent the end of this objectivism. But in what sense? This question shows that Rota, in the light of Husserl's reflection, aims to demonstrate "unequivocally the weakness of the idea of objectivity" (Rota, 1985e, p. 9), seeing it as *an obstacle* for the development of science. The impossibility of accounting for the phenomenon of sense on the basis of a world constituted by physical things is one of the clearest signs of the current crisis of science that, for Rota, manifests itself most strikingly in the studies on artificial intelligence. This discipline does nothing other than update—and enlarge—the tragic failure of modern *psychology* already denounced by Husserl (Husserl, 1959, p. 18).[12]

For Rota "the cleavage between science and philosophy in our time"[13] has complicated the development of those sciences, such as artificial intelligence, that involve questions that classically were considered to be exclusively under the dominion of philosophy. He maintains that philosophy must refrain from making predictions and limit itself, in its relationship with science, to a practice of "liberation from prejudice."[14] The nature of this and of other impediments connected with objectivism is historical, and it manifests itself fully in the epoch we are living through today. The end of objectivity thus assumes the connotation of a completion, of the exhaustion of a way that in the past was fertile and irreplaceable, but that is now no longer equal to those perspectives that it itself revealed.[15] This end regards exclusively the radical version of objectivism, because in its

---

[11] On the relationship between the conception of the transcendental in Kant and Husserl, see Derrida (1962), pp. 38–39: "Both the necessity to proceed from the fact of constituted science and the regression towards the nonempirical origins are at the same time conditions of possibility: such are, as we know, the imperatives of every transcendental philosophy faced with something like the history of mathematics. A fundamental difference remains, however, between Kant's intention and that of Husserl, one that is perhaps less easily distinguishable than would first be imagined."

[12] Marvin Minsky too, one of the founding fathers of artificial intelligence, endorses this close connection between artificial intelligence and psychological issues, but from a totally different perspective.

[13] Rota (1991a), p. 77.

[14] Ibid., p. 78.

[15] For an interpretation of the end as completion see Heidegger (1969), pp. 55–59.

non-dogmatic, "weakened" form, this attitude, conscious of its limits and of its history, is fully entitled to take its place within that scientific pluralism that Rota defends, following Husserl,[16] that "objectivity is just one project among many" (Rota, 1991a, pp. 124–125).

The constitution of "physical things" and the development of experimental techniques are obviously legitimate aspects of scientific practice, while phenomenology's accusations are directed against the claims of those scientists who embark on metaphysical speculations on the basis of "unrealistic philosophies of science,"[17] because

> [w]hen it is actually natural science that speaks, we listen gladly. But it is not always natural science that speaks when natural scientists are speaking; and it assuredly is *not* when they are talking about "philosophy of Nature" and "epistemology as a natural science." And, above all, it is not natural science that speaks when they try to make us believe that general truisms such as all axioms express [...] are indeed expressions of experiential matters of fact. (Husserl, 1913, p. 39)

2.2 ✧ Intentionality and Being-in-the-World

Rota is convinced that one of the fundamental tasks of phenomenology is that of highlighting the primordiality of sense. In his words, if "many disputes among philosophers are disputes about primordiality" then "phenomenology is yet another dispute about what is most primordial" (Rota, 1991a, p. 54). In this way he evidently does not intend to deny the existence of matter, of objects, or of that objective dimension proper to science, in favor of a spiritualist option, but rather to posit as primordial another dimension of the world connected with contexts and with roles, which is considered primordial because each one of us is confronted with it primordially. Following the road indicated by Husserl, Rota seeks to conquer a new perspective that, avoiding the blind alleys of objectivism and subjectivism alike,[18] acquires a new awareness of the problem of sense. To carry out this project he has recourse to the theory of intentionality that, as we know, represents one of the cornerstones of phenomenology.

Without entering into the details of an extremely complex debate, let me just say that it is important not to mistake the meaning of the word "intentionality" (as sometimes occurs in the analytic domain), lending it a juridical sense that assimilates it to teleology and interprets it as the plan of a subject who pursues an

---

[16] "In fact, we allow *no* authority to curtail our right to accept all kinds of intuition as equally valuable legitimating sources of cognition—not even the authority of 'modern natural science'" (Husserl, 1913, p. 39).

[17] Rota (1990a); see (1997a), p. 108.

[18] The second part of Husserl (1959), pp. 21–100, is devoted to an analysis of the opposition between physicalistic objectivism and transcendental subjectivism.

end. I shall deal with this question more thoroughly in the last chapter of this book, but in the meantime the following passage from Rota will permit us, provisionally, to clarify the meaning of the term.

> No object can be *given* without the intervention of a subject who exercises a selection of some features out of a potentially infinite variety. It is impossible, for example, to recognize that three pennies placed side-by-side with three marbles are both "the same number," without focusing upon "the number of" and disregarding other similarities [...] which may be more striking in other circumstances. This act of focusing [...] is *irreducible* to objectivity alone. It requires a contribution from the perceiving subject which, [...] is ultimately [...] *contingent*. [...] Yet, we know that objects come furnished with the quality of already-there-ness. [...] We are confronted with a paradox. On the one hand, the necessity of an observer turns the object into a contingent event [...]. On the other, pre-givenness of the world confronts us as the most obvious of realities. Like all paradoxes, the paradox results from withholding part of the truth, and it melts away as soon as the motivation for each of the two alternatives is looked into. (Rota, 1973b; see 1986a, pp. 247–248; emphasis in original)

Instead of constituting an unsolvable paradox, this alternative between necessity and contingency in the nature of an object reveals the woven threads of the fabric of intentionality. In fact, phenomenology develops an analysis of the "constitutive processes, [...] in the description of the sense of operations intentionally directed on multiple layers of the reality of the surrounding world" (Franzini, 1991, p. 25). In this way, for Sokolowski, Husserlian constitution is characterized as "the function of giving sense" (Sokolowski, 1964, p. 196). Unfortunately, the term "constitution" can also be misunderstood because it seems to refer to the function of an "abstract or intellectual 'I' that phenomenologically 'shapes' a world that is of itself shapeless and unknowable."[19] Obviously this is not the sense of Husserlian constitution, which is, rather, to be understood as "intentional description, i.e., as the giving of sense, which *recognizes* intentionally (and does not attribute 'from outside') the sense that lives in the multiple regions of *our* experience, and offers it *genesis*."[20]

The significance of the term "constitution" is one of the most complex questions of Husserl's philosophy, and one that can provide us with another coordinate to comprehend the philosophical references that are omitted in the work of "rewriting" represented by *The End of Objectivity*. Husserl's concept of constitution was the subject of a fundamental essay of Sokolowski's (Sokolowski, 1964) who, as we saw, was Rota's friend and colleague for many years and—we may well imagine—had a certain influence on him; and in fact there is evidence of his influence in some aspects of Rota's philosophical investigations on mathematics.[21]

---

[19] Franzini (1991), p. 26.
[20] Loc. cit.; emphasis in original.
[21] See Section 3.4.

Rota maintains that phenomenological analysis of constitution and of sense is essentially contextual.[22] Indeed, his particular attention to possible semantic misunderstandings led him to state that "there is a temptation for us to say that [...] sense is 'relative' to a context," but such an expression could implicitly suggest "a possibility for sense to be absolute,"[23] thereby surreptitiously smuggling objectivist prejudices back in. The specific meaning of the term "context" used by Rota can only be made clear in reference to the theme of the *world* (Rota, 1991a, p. 53)[24] that, in his rewriting, is interpreted as the context *par excellence* without which the phenomenon of sense could not be conceived.

Every genuinely phenomenological perspective must make its way through the insidious philosophical bottleneck to take its place with Husserl on a "neutral ground [...] this side of realism or idealism" (Vanni Rovighi, 1969, pp. 120–121). The world must therefore be conceived of in its equi-primordiality with the subject that "opens" it; and to this end Rota turns to Heidegger and maintains that our "basic relationship with the world [...] is [...] the relationship [...] with sense" that is made explicit in the modality of "being-in-the-world" (Rota, 1991a, pp. 61–62).

Rota's effort to tread a path free of the toll of objectivism leads him directly to one of the basic themes of Heideggerian phenomenology. In the objectivist tradition, as we saw earlier, our relationship with the world is forcedly passed through a sort of philosophical (or, if you prefer, methodological) sieve to separate what is primary in it, because it is attributable to the physics of the world of objects, from what belongs to the subject, and is thus secondary because it is mutable and contingent.

Objectivism, with its problematic correlate of the correspondence theory of truth, shatters the primordial phenomenon of being-in-the-world to attempt, subsequently, to build a bridge between its two isolated fragments, represented by the subject and by the world.[25] By contrast, Heidegger's *Dasein* ("being there"), as constitutively and primordially located in the world, cannot comprehend the world in an isolated manner since this operation would be equivalent to a separation from itself;[26] the one cannot be without the other since these two poles refer reciprocally to one another in their equiprimordiality. In this perspective it is no longer a necessity to refute idealism, understood as a problem of the proof of the existence of an external world independent of a subjectivity,[27] because "the question of whether there is a world at all and whether its Being can be proved, makes no sense if it is raised by *Dasein* as Being-in-the-world."[28]

---

[22] Rota (1991a), p. 126.

[23] Loc. cit.

[24] For an interesting analysis of the various interpretations of this theme in the phenomenological tradition, see Costa (2002d), pp. 258–263.

[25] See Heidegger (1927), pp. 252–254.

[26] Ibid., p. 186.

[27] Ibid., p. 247.

[28] Ibid., pp. 246–247.

At this point it is possible to reconsider (and to dissolve) the question regarding the origin of sense, from which we set out, by showing that it loses value if we relinquish physical things as a point of departure and foundation. With a sort of play of words it can be said that the question of the sense of sense is senseless because it pretends to surprise from a mythical "outside," with an objective gaze, the world, life, and our very subjectivity. This question must instead be turned upside-down, by showing that its genesis cannot be investigated on the basis of physical things; indeed, one must assume that "the world is [...] primarily a world of sense"[29] if one is to understand the significance of physical things. As Rota tells us, "phenomenology is [...] a description of the world—not of objects and things and physical laws, but of the world of that which we deal with."[30] This broaches a fundamental relationship of the worldly context that he describes as "dealing-with" (or "mattering"), which is his rendition of the Heideggerian term *Umgang*. Rota maintains that "dealing-with is the most neutral term [...] because dealing-with can be completely passive."[31] "Dealing-with" denotes "the basic relationship of man with world."[32]

The world is the primordial context but this does not mean that it is a unit, because contexts are not monads devoid of harmony that mutually exclude one another but are rather "layered upon each other in various ways."[33] Hence a phenomenon does not belong to just one of them but to a multiplicity from which, from one time to the next, the significant contextual sphere emerges. Rota often uses the word "role" in the phenomenological investigation of sense and its contextual characteristics because what we describe are not "things" "but only their 'roles' in contexts."[34] For him, the privilege attributed to things is due to "our prejudice that, in the layering of roles, there is some layer that is fundamental."[35] Nevertheless, also the term "role" seems inadequate because "unfortunately, the physicalistic language that we have inherited forces us to use terms that are very inappropriate." In this particular case "role" suggests the character of "a play as against the real person" while "our ultimate conclusion will be that there are only roles [...] and no 'real' life outside the play."[36] In spite of the inevitable compromise with objectivist language, Rota believes that the term "role" is preferable to others since it suggests contextuality.

Once the "worldliness" of sense has been attained, it has to be understood in its correlation with the things of the objectivist tradition. Hence Rota empha-

---

[29] Rota (1991a), p. 61.
[30] Ibid., pp. 1–2.
[31] Ibid., p. 55.
[32] Ibid., p. 62.
[33] Ibid., p. 126.
[34] Ibid., p. 124.
[35] Loc. cit.
[36] Ibid., p. 125.

sizes that one of the central problems is that "of the layering of the sense of our perception on underlying phenomena, and how this connection is made."[37] I shall return later, in discussing the relationship of *Fundierung*, to the question of the general characteristics of layering, but for the moment I limit myself to noting that the worldly phenomenon of sense is in relationship with a physical plane; for example, "my seeing something is layered upon certain perceptual phenomena such as my eyes and my brain."[38] Rota clearly shares Husserl's position that sense "is obtainable only in unity with natural and material phenomena" (Sini, 1987, p. 58). This observation is extremely important because it ought to defend Rota against the accusation of propagandizing a sort of ingenuous idealism or spiritualism.

This layering of the phenomena of sense with respect to a dimension distinguished by materiality is an issue that phenomenology will have to investigate. In fact Rota affirms that, in a phenomenological perspective, the world "is founded upon the world of 'matter' [...] but it itself, looked at nonreductionistically, is a world of contexts." The problem lies in understanding the term "to found" because, as we shall see, it refers to *Fundierung*, a Husserlian concept whose re-elaboration represents one of the most original elements of Rota's reflection. In his "discussion of Fundierung" Rota states, "We will also use certain English terms which have been used over the years as translations of Fundierung. They are layering, letting, and founding."[39] In Husserl (we shall examine the term in particular in the *Third Logical Investigation*) it is translated as "foundation," with the cognates "found," "founding," "foundational," "foundedness." In this regard, it is important to distinguish between the German terms *Fundierung* and *Begründung*, founding and grounding. While the former refers to a lower layer that founds a higher one (or ones), the latter refers to the "ground," to that which founds (or "grounds") without itself being founded ("grounded"). Grounding is similar to the building of a basement that would support the whole building; founding is similar to the building of a floor that would support only the floors above.

Thus *Fundierung* indicates a relationship of dependency between various layers of the world that can be interpreted in either a reductionist or a phenomenological sense. In the first case, it indicates the reduction of an apparent layer to a real one of which the former is its epiphenomenon. This is not the meaning of *Fundierung* as foundation that Rota begins to clarify with a series of examples that must be patiently followed if one is to trace the comprehensive meaning of his reflection on this theme. As we have seen in the first part of this book, the use of examples is a distinctive characteristic of his modality of investigation that, moreover, provides us with important information on his sources. The first example, to which

---

[37] Ibid., p. 53.

[38] Loc. cit. In this regard Merleau-Ponty states that "the relationship between matter and form is called in phenomenological terminology a relationship of *Fundierung* [foundation]: the symbolic function rests on the visual as on a ground" (Merleau-Ponty, 1945, p. 146).

[39] Ibid., p. 42. In *The End of Objectivity* "Fundierung" is not italicized.

I shall frequently refer, concerns the game of bridge.[40] If I pick up a playing card asking someone whether it is a trump[41] I will receive an answer that depends on the context. The trump card is not an arbitrary phenomenon, since any observer who knows the rules of bridge and the development of the game is able to answer univocally; but it is not a "fact" either, since this situation cannot be grasped from an analysis of the pure physical characteristics of the cards. In this way that which common language calls "fact" is transformed, in the phenomenological perspective, into a "contextual happening."[42] In any event, we still have the open question of the relationship of sense with that dimension of the "physical thing" relevant to objects and to facts that Rota calls *facticity*, we shall investigate it in depth later on. For the moment I shall dwell upon the *pars destruens* of his reflection aimed at a distancing from the objectivism with which common sense and scientistic philosophies are imbued.

To conclude this part of my analysis I have to come to terms with another aspect of objectivism criticized by Rota: reductionism. In his view "the main thesis of phenomenology is that there is a non-reductionist attitude we can take towards every human phenomenon, and we can only understand the phenomenon if we describe it 'as such,' previous to any reduction to other attitudes."[43] In this perspective it can be said that for any given phenomenon there exists a non-reductionist attitude that not only possesses equal dignity with respect to the reductionist attitude but that also must precede it because it is primordial. In fact one must never forget that science is "a human spiritual accomplishment which presupposes as its point of departure, both historically and for each new student, the intuitive surrounding world of life, pregiven as existing for all in common" (Husserl, 1959, p. 121).

Rota criticizes a great number of philosophical and scientific opinions for their reductionist approach that he ascribes to a materialist (or perhaps more precisely, "physicalist")[44] attitude, and that he describes as an interpretation of reality according to which really existing things manifest themselves physically.[45] Rota defends the autonomy of every level of description and criticizes the attempts to reduce music, psychology, or biology to sophisticated corollaries of physics, since every worldly phenomenon can be genuinely understood only if one grasps its intrinsic, primordial, and irreducible sense.

This identification of reductionism with materialism—so blatantly unilateral and biased—was undoubtedly influenced by the philosophical opinions dominant

---

[40] Ryle (1954) inspired this example.

[41] A suit that in briscola is determined by picking the top card of a just-shuffled deck, and in bridge by means of a process of "bidding" among the players at the beginning of the game.

[42] Rota (1991a), p. 10.

[43] Ibid., p. 75.

[44] Rota (1991a), p. 25.

[45] Loc. cit. For Rota, one of the most traditional expressions of this prejudice is represented by the mechanicism that is presented as the classical form of materialism; ibid., p. 138.

in the English-language cultural context that Karl Popper (1902–1994) critiqued.[46] I believe that this imbalance within Rota's argumentations was part of a rhetorical strategy that aimed to fortify the side of the wall that was exposed to the attacks of his most numerous and formidable enemies.[47] Leaving these rhetorical and cultural aspects aside, and to understand better the successive developments of his phenomenological investigation, I propose to analyze reductionism in a different perspective, interpreting it, in mereological terms, as

> the opinion according to which any property of a complex system (a whole) is completely determined by the structure of the system, i.e., by the number, characteristics, and mutual relations of its constituent parts [...] [T]he reductionist [...] will maintain that the properties of wholes can be explained [...] on the basis of their structure, and it is precisely this explanation that is called reduction [...] [According to the radical reductionist] in the whole there is nothing more than the parts of which it is constituted [...] the reductionist [...] considers only the determination of wholes by means of their parts. [The reductionist] does not take into consideration the possibility that an object, becoming a part of a complex structure, can be modified, [or] that the properties it manifests independently of the structure may not always be the same [properties] it manifests when it becomes a part [of that structure]. (Amsterdamski, 1981, pp. 63–64)

In this perspective I intend to restructure the sense of Rota's statements, showing how the end of objectivity and the critique of reductionism are founded upon the complex relationship between the whole and the parts and on the relationship of *Fundierung*. The latter is the true keystone of his critique of reductionism, since it allows us to understand the manifestation of "emergent novelties" in the whole that are absolutely not reducible to the resultant of the simple parts.

Fundierung will clarify both the critique of reductionism and the possibility of a correlation between sense and physical things. To develop this interpretation fully we will have to take a few steps beyond Rota's text to grasp the importance that Husserl's theory "of the wholes and of the parts" possesses—with regard to the *Fundierung* relationship—for the critique of reductionism. Since Rota himself wrote but a single text on this specific theme, *Fundierung as a Logical Concept* (Rota, 1989a),[48] we shall have to track down the relevant references in a number of essays and articles in which he disseminated his ideas. The description of *Fundierung* entails "a difficulty similar to that of the logician who teaches the basic operations between sets [...] A teacher has

---

[46] Popper criticized the fact that, for the first time, the absolute denial of the existence of consciousness entered the universities, in support of a radical version of fashionable materialism. See Popper (1991).

[47] "Seventy-two years after the publication of Husserl's *Logical Investigations*, the barbarians are besieging the citadel of phenomenology and crying in guttural accents: 'Put up or shut up'" (Rota, 1973b; see 1986a, p. 252).

[48] See Rota (1997a), pp. 172–181.

no alternative but to proceed indirectly, leading his students through a sequence of examples, hoping that the underlying concept will eventually shine through."[49] Paraphrasing Heidegger we could say that from the question "what is *Fundierung?*" one awaits a definition and a technical answer, which Rota foregoes to examine some exemplary relationships. I shall begin my exposition with the relationship he considered one of the most important: the phenomenon of reading.

2.3 ✧ Reading and Seeing

Each morning we open our eyes upon a world covered with notices, indications, articles, prohibitions, captions, and road signs that follow us throughout the day. Paraphrasing Derrida, it could be said that we are always watched over by a text. The function of reading is so pervasive in our society that it is difficult for a completely illiterate person not only to work but, simply, to survive without the ability to distinguish the label on a bottle of mineral water from the one on a bottle of acid. We all belong to a world of letters and words in which reading appears as the most obvious of human activities—and yet this familiarity does not mean that we understand what reading is. Not even the scientific disciplines can help us to grasp definitively this phenomenon of which we have only an average and vague understanding, because behind the obviousness of this "simple act" lurks the problem of sense. This is why philosophy in this epoch needs to rethink and rediscover the sense of such simple actions as seeing and reading.[50]

Rota denounces the insufficiency of the reductionist interpretations of reading[51] that simplify the phenomenon, eliminating all its characteristics that are considered extraneous. A typical example is the attempt to derive reading from the physical act of looking at ink stains that subsequently are interpreted as symbols having meaning.[52] In this direction Rota rereads[53] some interesting analyses of the phenomenon of reading contained in the *Philosophical Investigations*,[54] going well beyond the intentions of Wittgenstein who explicitly declares that he is "not counting the understanding of what is read as part of 'reading' for purposes of this investigation."[55] Nevertheless, Rota insists that two meanings of the term "reading" are present in Wittgenstein: the first refers to physical (neurological, biological, and optical) processes, while the second regards function.

It is necessary to make explicit and to develop Rota's reflections in order to show how the reduction of reading to the interpretation of ink stains, understood

---

[49] Ibid., p. 173.
[50] See Althusser and Balibar (1965), p. 16. See also D'Alessandro (1980).
[51] See Rota (1991a), p. 39.
[52] Loc. cit.
[53] Rota (1985e), pp. 8–9.
[54] Wittgenstein (1953), pp. 61–70.
[55] Ibid., p. 61.

dogmatically, leads to aporetic outcomes that deny aspects of that selfsame material reality to which everything is intended to be referred. In fact, the conviction of the priority of physical existence leads one to deny (paradoxically) the existence of many "real" and everyday situations, such as this sheet of paper on which you are reading. In essence, Rota was attempting to reassert Husserl's criticism of the empiricists who "have apparently failed to see that the very scientific demands that they, in their theses, present to all cognitions are also addressed to those theses themselves" (Husserl, 1913, p. 38).

For Rota, materialist reductionism claims that the understanding inherent in the reading of any writing—the page of a newspaper, for example—derives directly from the atoms and the molecules of the paper and the ink that "mysteriously organize into the copy that becomes the newspaper" (Rota, 1991a, p. 65).[56] If reductionists were consistent they would have to admit that this mysterious organizing, even if unknown, will itself have to be reducible to physical reality. By granting reality only to matter (in an exaggerated and extreme way) one would have to conclude, in a manner that is evidently absurd, that the phenomenon of the understanding of these lines of text does not exist; in fact no chemical or physical analysis of a sheet of paper can render in itself and for itself the informatory content of a written page. Taking a very simple example, let us consider the two words "philosophy" and "PHILOSOPHY."[57]

Physically the two words have very little in common. The first is noticeably smaller than the second, and consequently a smaller quantity of ink was employed for it. Furthermore, let us suppose that the first is written with red ink on parchment while the second is reproduced on a computer monitor. Comparing in the same order the letters that compose the two words (the shape of the ink stains that constitute the letters of the first and the pixels of the second) we realize that they are significantly different. At this point, following a purely physicalist logic we could not conclude that the two words have the same meaning, because the rules of reading transcend the materiality of what is written and physically do not exist anywhere.

The objection that calls upon manuals, dictionaries, and texts of phonetics containing rules and indications is ineffective; for a radical materialist such rules are a strong-box that is irremediably closed due to the paradox triggered by "physicalist dogmatism." In fact, to learn to read one must read the dictionary; but to be able to read the dictionary one must *already* be able to read. The rules of reading contain a sense that cannot be disclosed precisely because of this philosophical obstinacy:

---

[56] In particular, Rota criticizes the positions taken by the neurologist Steven Rose; see Rota (1997h), p. 115 (republished in 1997a, p. 187). Positions of this type are widespread among analytic philosophers; in this regard see Fodor (1975) and Dennett (1991).

[57] This argumentation was developed based on my reading of Rota's texts; I subsequently had occasion to discuss it with him at great length.

for an extreme materialist, the rules of literacy are literally metaphysical. Obviously this is a banal example and is effective only in the case of an ingenuous and caricature-like reductionism; but, nevertheless, even the most sophisticated reductionist hypothesis is based on the unexpressed precomprehension of the identity of the meaning of two letters written in significantly different ways. This is the fundamental situation I am anxious to make clear. We can say that learning to read means facing the circular structure of comprehension that Heidegger called the hermeneutical circle.

Furthermore, if we analyze the problem not simply at the lexical level but go on to the levels of grammar and syntax we become aware of the difficulty involved in segmenting the comprehension of a written text into a succession of logical steps or of algorithms. This is because the sense of a sentence is, in general, not grasped in a cumulative way, and (inversely) not in a series of successive approximations either. Despite the apparent simplicity of our modality of writing, which proceeds from right to left and from top to bottom and uses spacing to subdivide groups of letters into words, the comprehension of a text cannot be reduced to a linear succession. We are, rather, confronted with a complex phenomenon in which sense is not only accumulated (or approximated) linearly and unidirectionally but is also layered in a multilinear and multidirectional structure. The new words that we read towards the right can also invest with new meanings some of the words we have already read on our left.[58] This is why the "notion of textuality is that which is not given in an immediate way to what is intuitively evident, but is constituted in a *reference*."[59] In fact we must never forget that "the word of a text does not 'mean' as such, but in relation to other, to something that is not present, but that intervenes in the very heart of the word to constitute it."[60]

This phenomenon is generalized if we think of the relation between sentences, and of internal references between pages, chapters, or books. Here, we are confronted with the extraordinary complexity of the phenomenon of reading that different theories have had to deal with, as in the case of systems theory with its concept of feedback, or of hermeneutics with the circular relation between text and context (to limit ourselves to two approaches that Rota knew very well). Albeit from different perspectives, both approaches highlight that problem of the relation between whole and part in the phenomenon of reading which represents, as we shall see in a mo-

---

[58] Lacan examined this phenomenon in the structure of the *après coup*, showing that the close of a sentence has a retroactive effect of sense; see Lacan (1966), p. 200. In this regard Rota told me that he had been the witness of an interesting episode concerning the French psychoanalyst. His account helped me to clarify Lacan's interest in topology and to structure the central part of the monograph I have recently dedicated to this scholar; see Palombi (2009b), pp. 43–48 and pp. 111–112.

[59] Dalmasso (1990b), p. 49.

[60] Loc. cit.

ment, a crucial element also for an understanding of Rota's thinking, and which connects his critique of reductionism with his reflection on *Fundierung*.

These considerations also allow us a digression on the theme of hypertexts, whose applications have been rather ubiquitous in recent years. Normally, the most common approach to the hypertext is that of describing it in opposition to the linear text, but what we have just seen shows that, strictly speaking, there are no linear texts and *all* texts are hypertexts. From a technological standpoint, the novelty of the advances in informatics consists in an extraordinary increase in the speed of managing information, while from the philosophical (and didactic) standpoint it consists in the possibility of showing the functioning "in action" of *every* text.

At this point, if we are to avoid the aforementioned absurdities, we will necessarily have to conclude that the phenomenon of reading cannot be deduced from the simple materiality of what is written. This situation involves not only the understanding of a text but also the more general understanding of what occurs in phenomena of perception, since analogous considerations can be developed also for the figures of Gestalt. In fact, if we persist in granting reality to physical existence alone we absolutely cannot understand the extraordinary change that has taken place in the meaning of a figure that has remained physically unchanged. Rota further considered the game of chess in his criticism of reductionist aporias (probably inspired by Wittgenstein, 1953, p. 15). Although this game requires pieces and a chessboard, Rota emphasizes that we can understand nothing about its rules from the simple observation of the pieces and board themselves, which can be of very different shapes and materials. Furthermore, he notes that the mental acts required for playing "do not relate" to the rules of the game "except to someone who [...] *already* knows chess" (Rota, 1991a, p. 41; emphasis added). This example is a variant that presents the same argumentative structure as the example of reading but that is not able to refute the potential objection that calls for a manual for the game of chess. In fact, in the case of reading, every manual becomes inaccessible since it is incomprehensible. The example of reading, if compared to that of chess, seems to drive the reductionist and the radical materialist into a sort of blind alley.

After this analysis of the paradoxes produced by the identification of reading with physical process alone, let us return to analyzing it from the standpoint of function. Rota tells us that "The text is the facticity. The content of the text is the function, in this case, the sense." He affirms that, for any text, what matters to us is its content; if we identify sense with materiality we are in error.[61] In fact it is easy to imagine that the *same sense* can be summarized in another text, repeated orally, or circulated by email on the Internet by means of a modem.

Rota wonders about this erroneous identification, but above all about its condition of possibility represented by the indispensability of a physical substrate that

---

[61] See Rota (1991a), p. 67.

leads to confusion between a text and its content.[62] Hence "logical hygiene demands that we keep the terms 'text' and 'content of the text' separate and equal. The text may be an object. The content of the text is not an object in any ordinary sense" (Rota, 1989a; see 1997a, p. 174). One must therefore conclude that content—sense—cannot be located in any physical place.[63] If you sought, for example, to get out of the difficulty by locating it in the mind, you would have to face the even more difficult problem of "the relationship between the content and your mind," which cannot be identified without "making a crass reduction or error."[64]

Nevertheless, it must be emphasized that even though the phenomenon of reading is correlated to the understanding of a *quid* (a sense) that is not physical, Rota is by no means a supporter of any sort of spiritualism or of an updated version of occasionalism. He seeks to tread a sort of philosophical third way that takes into account the materiality of the text and the physical elements of reading, but insists that sense cannot be reduced to these elements.

We can better understand this effort, which represents one of the hallmarks of his thinking, if we consider his reformulation of the question that correlates the text with the facticity and the sense (the content) with the function, and that reflects on the relationship of dependency that subsists between these two levels (Rota, 1989a).[65] The first characteristic feature of this relationship is that "it is not transitive,"[66] which makes it impossible to reconstruct a chain of cause and effect that, considering the dependency of sense on the reader's behavior, and the further dependency of sense on the text's materiality, would refer and reduce everything to the physical level. This brings into focus one of the fundamental problems of Rota's phenomenological analysis: "the problem of understanding what is to be meant by *dependency* of the content of what I read on the text that I read."[67] Rota describes this relation, exemplified in the phenomenon of reading by the relationship between text and sense, in terms of Husserl's notion of *Fundierung*, which we shall now investigate more closely.

2.4 ✧ *Fundierung* in the *Third Logical Investigation*

To investigate the relationship of *Fundierung* I shall return to Rota's analysis of contextual sense touched upon earlier. His remarks are evidently inspired by the texts of Husserl and Heidegger even if, consistent with Rota's practice of rewriting, these authors are not always explicitly mentioned. First, Rota poses the lexical

---

[62] Loc. cit.

[63] Loc. cit.

[64] Loc. cit.

[65] See Rota (1997a), pp. 174–181.

[66] Ibid., p. 175.

[67] Loc. cit.

question regarding the necessity of indicating the elements present in a context correctly, while simultaneously avoiding the term "object," in order to avoid any inappropriate reference to the physical dimension.[68]

This precaution, however, is not always consistently respected, and at times the term manages to sneak in, complicating the understanding of the passage. In any event Rota, basing himself on the assumption that the sense of an element of a context is "to function for some purpose," uses the term "function" to indicate that which might otherwise be called "contextual object."[69]

In order to clarify *Fundierung* in the light of function, I shall examine an example that Rota extrapolated from Ryle's book *Dilemmas* (Ryle, 1954), and that he subsequently re-elaborated from the standpoint of his phenomenological perspective. Every card of a deck can be used for countless different games. Hence, for Rota,

> [t]here is a relation of *Fundierung* between the *function* of the queen of hearts, whether in poker or in bridge, and the actual card. One cannot infer the function (the "role") of the queen of hearts in either game from plain knowledge, no matter how detailed, of the queen of hearts as a card. (Rota, 1989a)[70]

Consistent with what we have said about the phenomenon of reading, we must note the senselessness of the question regarding the physical place in which one can situate the role of the queen of hearts in a game (of bridge, for example). Rota emphasizes how such a role is correlated by complex *Fundierung* relationships to "brain processes, to the physics of playing cards, to the players' ambiance, and so on, *ad infinitum*,"[71] and yet the role itself is not reducible and is not identifiable with any physical phenomenon. This example highlights a fundamental aspect of the theory of *Fundierung*, namely: "When we focus on the context of a text, or on the role of the queen of hearts in a bridge game, we focus on a contextual *function* of the content. The *function* of the queen of hearts in a specific bridge game *matters*."[72]

Rota conceives of *Fundierung* as a relationship constituted by two correlated terms: facticity and function. Facticity (represented in our previous examples by the written text, or by the queen of hearts as a mere card of the deck) plays a role in sustaining function, but only function has a strict relation with sense. Facticity represents an indispensable element of access to function; in fact, it is impossible to play cards without some type of factic support upon which to represent the cards (whether cardboard or a computer monitor), and yet one cannot individuate in facticity the sense of any game. These considerations hold true more generally for every form of text or language.

---

[68] See Rota (1991a), p. 157.
[69] Ibid., pp. 126–127.
[70] See Rota (1997a), p. 175.
[71] Loc. cit.
[72] Ibid., p. 176.

What, in essence, is *Fundierung*? I could say that

> [a]s I speak to you, you listen to me through sounds which somehow make sense. Sense and sound are related by a *Fundierungszusammenhang*, a relation that founds. The sense of my words cannot be without the sound, but I cannot glean the sense from the simple sound as such [...]. Husserl gave profound and very beautiful descriptions of *Fundierungszusammenhang* and of its variants, the upshot of which is to prove that the expression "A is founded on B" is independent and autonomous of the relations of membership and containment. (Rota, 1987a, p. 124; see 1997a, p. 190)

Fundierung is described as a primordial relationship that is irreducible to more simple relationships (material relationships in particular) because otherwise we would confuse function with facticity, falling into a typical reductionist error. Facticity is necessary for the relation of foundation because if we eliminate function, facticity will vanish along with it. Nevertheless facticity is not self-sufficient and must not overshadow the function it founds. Rota emphasizes how this tenuous umbilical cord between function and facticity is a source of anxiety for reductionism and for common sense (considered as a sort of unconscious reductionism) that wish to identify functions with physical objects. I recall, for example, that both reductionism and common sense harbor the illusion of being able to reduce functions to simple brain processes, thanks to psychology and the cognitive sciences, clinging desperately to anything that can free them from the burden of an autonomous and non-reductionist theory of *Fundierung*. Rota maintains that it is possible to subsume the concept of role, previously broached, under the more general concept of function, as the following example of the passenger shows.[73]

Imagine you are in an airport in front of a door that leads to the runway. A plane lands and a small crowd of persons comes out of the door and is immediately identified by the passport control officer as a group of incoming passengers. Now, suppose that the same scene unfolds in an unspecified place. In this second case the officer would not be able to interpret as passengers the persons who swarm out of just any door, since they might equally as well be clerks leaving an office, or soldiers in civilian dress who are going on leave, as travelers arriving from abroad. This example ought to highlight the fact that one can identify a class of persons on the basis of the role they play against the background of a given context. Hence Rota concludes that "in no case do we have a pure and simple perception of persons or of things without a simultaneous identification of a function."[74]

The example of the passenger is to be interpreted on the basis of the relationship of *Fundierung*, illustrating the fact that roles, too, have need of a factic and a functional pole. We identify the factic in the persons' bodies, in the fabric their

---

[73] Rota (1985e), p. 9.
[74] Loc. cit.

baggage and clothing is made of, in the direction in which they walk, in the paper their tickets are printed on, and in the airport structures. Onto this (obviously partial) list is grafted a series of intentional references that correlate, for example, the paper and the ink of the tickets, to the places the passengers are coming from, to the writing on the arrivals board. All of these are functions standing out against the contextual background represented by the airport.

This contextual aspect is fundamental also for understanding the importance Rota attributes to the "tool," seen as the fundamental demand of *Fundierung*, and to its context represented by the "tool room" (Rota, 1991a, p. 157).[75] Imagine a tool room in which you are surrounded by objects that are called tools because they are meaningful in light of a possible use, while this definition is perfectly indifferent to their chemical and physical composition.[76] Why is it that in a tool room a hammer and a nail seem familiar and appropriate while a comb appears out of place? In a horizon characterized by objectivism one ought to start out from the nail and the hammer to discuss the conditions of possibility of their reciprocal correlation and of their being correlated with the tool room. For Rota such a reflection goes in the wrong direction and therefore must be reversed; it is not possible to conceive of the factic and separate existence of the two tools as the condition of possibility of their correlation but, on the contrary, it is precisely the relation that represents the transcendental dimension of the nail and of the hammer and, in general, of the tool room.[77] In the same way, the out-of-placeness of the comb must be grasped on the basis of the contextual background that does not disclose relations of usability with the background of the tool room. Conversely, in a barbershop the presence of a hammer would be equally enigmatic if in the context there were not something to break or hit.

To pose the question about the relational possibility of objects of common sense is erroneous: "When we ask the question how do they relate, we are lying, because we do not first find the object and then ask how we relate it. What you find from the start are objects that are already related to each other by their possible joint functionalities" (Rota, 1991a, p. 117).[78] Rota affirms that the characteristic feature of "objects in the context" is precisely that of being "generalized tools—namely, functions."[79]

Functionality can be properly described only as "part and parcel of a network of referential functions of possibilities"; just as "you cannot define trump card

---

[75] See Heidegger (1927), p. 105; and Wittgenstein (1953), pp. 6–7. The examples are inspired by Heidegger, as Rota states explicitly (see Rota 1991a, p. 157, note 24). I must, however, note that Rota uses "tool" as the translation of the German *Zeug*, while in fact the English translation of *Sein und Zeit* uses "tool" to translate *Werkzeug* (Macquarrie and Robinson translation, p. 575). Furthermore, Rota translates the German *Werkstatt* to "tool room," instead of the term "workshop" used in the English translation.

[76] Gian-Carlo Rota, conversation with the author; see Heidegger (1927), pp. 96–97.

[77] See Heidegger (1927), pp. 96–98.

[78] Ibid., p. 97.

[79] Rota (1991a), pp. 126–127.

alone without referring indirectly to the entire game of bridge," so "we view the tools as possibly being correlated to an indirect function."[80] From the phenomenological viewpoint, pencil, paper, and ink are not physical things but *tools*, i.e., "*functions* in *Fundierung* relations."[81] In fact, the pen with which I write, commonly considered from a material point of view, is, strictly speaking, a "*function* that *lets* me write."[82] Rota maintains that we can recognize a certain tool as such only by virtue of our familiarity with its possible functions. This is why he criticizes the absurdity of the physical reductionism that, examining the simple assemblage of plastic, metal, and ink, will not be able to grasp the nature of the pen, which is accessible only to one who wishes to see its function beyond all the facticities upon which it is founded. This example ought to make it clear that

> [a] craftsman, a scientist, a man of the world regard the objects of daily commerce primarily as *tools*. It is a prerequisite of a tool that it offer its user a reliable guarantee of sameness. Only when the tool fails or becomes otherwise *problematic*, as when the need for a new tool is felt, does the contingency of all toolness reveal itself to the user, like a crack in a smoothly plastered wall. In such moments of *crisis*, the tool-user is forced to engage in a search for origins, for lost motivation, for forgotten mechanisms that had come to be taken for granted, much like a garage mechanic in his daily work. (Rota, 1973b)[83]

In this passage, Rota reformulates Heidegger's thesis that individuates in usability "the primordial dimension of the relationship with a being within-the-world, which is not primarily known in its being in itself, but 'read' [...] according to contextual references."[84] Hence it is important to reassert that only subsequently can "the object isolated as *thing*" derive from the "item of equipment [*Zeug*] when [...] its functionality to the environment [...] enters in crisis."[85] In the last lines of the passage above, Rota emphasizes the importance of situations of crisis that, as we shall see, take on fundamental importance in his philosophical analysis of mathematical research. Such research always keeps in mind the function of phenomenological *epoché*, or "bracketing," reinterpreted as the effort to "isolate the purely contextual component of functions" in order to describe this component independently of facticity (Rota, 1991a, p. 143). Following the example of Heidegger's thought,[86] one

---

[80] Ibid., p. 118.
[81] Rota (1989a); see (1997a), p. 177 (emphasis in original).
[82] Loc. cit. (emphasis in original).
[83] See Rota (1986a), p. 248 (emphasis in original). This passage is clearly inspired by Heidegger (1927), p. 103.
We recall, moreover, that also Kuhn affirms: "As in manufacture so in science—retooling is an extravagance to be reserved for the occasion that demands it. The significance of crises is the indication they provide that an occasion for retooling has arrived" (Kuhn, 1962, p. 76).
[84] Marini (1982b), p. LVIII.
[85] Loc. cit.
[86] See Heidegger (1927), pp. 102–103.

cannot forget that situations of crisis can also be reinterpreted contextually, since among the numerous ways of being in relation in a context there is also the possibility of dubiousness, obtrusiveness, and irrelevance.[87]

At the end of this analysis, following Rota we can adopt the expression "network of functions" to indicate the contextual elements. Thus Rota concludes that "the objects of everyday commerce are really just pure functions and there are complex Fundierung relations that relate them to physicalities." In this same perspective the world is disclosed to us as a "network of related functions."[88]

In order to investigate more thoroughly this disclosure of the world in a phenomenological perspective we must now examine *Fundierung* in relation to Husserl's theory of the whole and to the example of the star, which represent the hallmark of my present work.

2.5 ✧ The Star, the Whole, and the Part

I defined reductionism with regard to the relationship between whole and part to overcome the limit of Rota's view that seemed to identify it with physicalist materialism.[89] Furthermore, I proposed to interpret his critique in the light of his analyses of Husserl's concept of *Fundierung* in the *Third Investigation*, which, significantly, is dedicated to the theory of wholes and parts. At this point I believe that all the elements are available to show that Rota does not bring to completion an important aspect of his reflection,[90] but limits himself to a critique of physicalism and of mechanicism without opening up a broader philosophical horizon.

I want first to emphasize that the meaning Husserl attaches to *part*, defined in the *Third Logical Investigation* and shared by Rota, is very far from the one normally accepted in common language. In fact, Husserl states: "We interpret the word 'part' in the *widest* sense: we may call anything a 'part' that can be distinguished 'in' an object." In this enlarged meaning, "everything is a part that is an object's real possession, not only in the sense of being a real thing, but also in the sense of being something really in something, that truly helps to make it up." It is evident that "the term 'part' is not used so widely in ordinary discourse."[91] The extremely broad range of this conception can be understood better if we consider the fact that, in this acception, predicates such as "red" and "round" are also admitted as parts.[92]

Husserl's perspective is revolutionary because it admits as parts of the whole even characteristics that entail pure possibility, in an effort to conceive of an extraordinary variety of possible parts that have nothing else in common apart from

---

[87] Rota (1991a), p. 143.
[88] Ibid., pp. 158–160.
[89] See Section 2.1.
[90] See Cerasoli (1999), p. 14. Moreover, Rota himself expressed appreciation for these observations; see Rota (1998f).
[91] Husserl (1900–1901), Vol. II, pp. 4–5.
[92] Loc. cit.

belonging to an object in a broad and extremely far-reaching sense.[93] Husserl speci-
fies his concept on the basis of a "cardinal difference" between independent and
non-independent parts, respectively defined as "pieces" and "moments."[94] I do not
wish to enter into the merits of the phenomenological analyses developed in the
Third Investigation, because I do not presume to appraise the different interpretations
of Husserl's text but, rather, only to use it, as I have done until now, to understand
Rota's philosophical perspective.

So, my point is this: in Husserl's text the notion of "whole" is instituted "on
the basis of the relation of non-independence, which receives [...] the denomina-
tion of relation of foundation" that is understood as "primordial" (Piana 1977,
p. 19). We thus begin to clarify the relationship between the theory of wholes and
parts and the relation of *Fundierung*, whose particular relation of *dependency* Rota
emphasized. In fact, the notion of "whole," defined by way of the concept of
Fundierung, refers to "a range of contents which are all covered *by a single foundation* without the help of further contents. The contents of such a range we call its
parts" (Husserl, 1900–1901, Vol. II, p. 34).

I have often asserted that Rota does not contest the results of objectivist and
reductionist science or (obviously) its right to pursue its own research perspective;
what he criticizes is the pretense of objectivism and radical reductionism to claim
to be the only legitimate philosophical interpretations of science. Since I have connected two different conceptions of the relation between whole and part with these
different perspectives it is necessary, from a pluralistic standpoint, not to deny one
or the other but, rather, to delimit their respective fields of validity. The possibility
of mapping out this delimitation manifests itself clearly by pitting

> the term "whole" (implying its "organicity") against the expression of "mere
> sum" (indicating an ensemble analyzable in its parts in the same way that a wall
> could be materially disassembled into the bricks it is built of) [...]. In fact, the no-
> tion of "whole" determined on the basis of the relation of foundation draws a
> distinction [...] between ensembles of objects that are whole [...] and ensembles
> that are not [...]. Hence just any given ensemble of objects [...] will not be said to
> be whole. Ensembles of objects that are not wholes will be [...] characterized by
> the fact that they are not covered by a unitary foundation. If we decide to reserve
> the term "set" for the latter, then we shall distinguish wholes from sets. (Piana,
> 1977, pp. 24–25)

For Rota, Husserl's concept of whole in the pregnant sense (*der prägnante Begriff
des Ganzen*) is precisely intended to distinguish the whole as function understood in
a relation of *Fundierung* from the set as a neutral collection of heterogeneous parts.
Rota, however, (once again) declares himself to be dissatisfied with Husserl's terminology, because it is still easily equivocal. Husserl defines the relation of *Fundierung*

---

[93] Loc. cit.
[94] Loc. cit.

as a particular relation of dependency between two "poles," the "moment" and the "whole," in order to emphasize the distinction between whole and set more effectively. Rota calls the first of these two poles "factic moment" or "facticity" and the second, "function." We can thus affirm that between whole and set lies the border between a non-radical reductionism and the phenomenological perspective that Rota intends to evidence and to valorize.

We need to examine the discourse more closely by confronting the problem of the layering of the relation of layered foundation, of which Husserl emphasizes the singleness insofar as *"every content is foundationally connected, whether directly or indirectly, with every content"* (Husserl, 1900–1901, p. 34; emphasis in original). It is therefore possible to distinguish between parts that are in a relation either of mediated or of immediate *Fundierung*, because "the iterability [...] of the relation of mediacy permits the introduction of differences of degree," thus making it possible to individuate a "greater or less 'closeness' [...] of the parts with respect to the whole" (Piana, 1977, p. 29). These differences of degree are defined by Rota as layerings.

To understand Rota's analysis, and to exemplify his conception of a layered relationship of *Fundierung*, let us consider Husserl's key example of the star-shape.[95] He writes of a complex star-shape built out of smaller star-shapes, which in their turn are composed of stretches, built out of segments that are, ultimately, made of points. In the phenomenological perspective we can say that there is a first relation of *Fundierung* between the points and the stretches and a second, layered upon the first, between the stretches and the star. Since we see the star through the stretches and the stretches through the points, in Rota's terminology we can say that the stretches are factic moments of the star and the points are factic moments of each stretch.

The *whole* that Husserl describes as "star-shape" is translated in Rota's vocabulary by the term "function." Stretches, segments, points, and other characteristics regarding this shape in any sense are admitted as parts of the same "star-shape." In the example the star represents a whole in the phenomenological acceptation, which, as function, is founded both by the stretch-parts and by the point-parts. The points, segments, stretches, and star-shape have "a fixed order of 'foundations,' in which what is founded at one level serves to 'found' the level next above, and in such a manner that at each level new forms, only reachable at that level, are involved."[96]

For Rota, in a layered sequence of relations of *Fundierung*, which correlate the parts of a succession A, B, C... to one another, term B will be, respectively, the factic moment of term C and the function of term A in a layered relationship of *Fundierung*. Each of these layers depends on the successive layer without being determined by it: *"Fundierung* [...] is a new kind of dependency which we need to

---

[95] Husserl (1900–1901), p. 40.
[96] Loc. cit.

properly describe without presuming that one layer determines the other." Hence Rota insists that "the layers have to be kept separate and distinct, they have to be described independently."[97]

Thus Rota, in a personal way, re-elaborates the relation of *Fundierung* while passing over Husserl's argumentations; hence I will limit myself to presenting his theses without further reference to Husserl's text. In this extremely general and broad sense of the concept of the *Fundierung* relation, a hypothetical list of the parts of a word written on a sheet of paper ought to include not only the dots of ink, the lines that trace every letter of the word, and the letters of the word itself, but also the language in which the word acquires sense, the contextual sense of the word, the senses in which the word is understood, the possible meanings the word can acquire in other contexts, and so on.

Going back to the example of a word written on a sheet of paper, Rota maintains that when the phenomenological list of the parts grows longer, we can see a superposition of foundations that constitute a layering wholly analogous to the one examined in the example of the star-shape. In general, then, it can be said that, in a whole, two layers that present themselves as successive constitute a couple whose elements are facticity and function. For example, the parts "letters" and "word" belong to two successive layers, and "word" and its "sense" to two others (omitting momentarily the problem of polysemy). It is impossible to characterize with precision the notion of "successive layer" and to determine further the relation between the letters that constitute a word and the word itself, apart from recognizing such a relation as an instance of *Fundierung*.

The layering of the relations of foundation can be applied to Ryle's example, imagining that in a certain game the trump card is the queen of hearts. In this way one emphasizes that the phenomenon of a given card of the deck being a trump is founded upon "a sequence of 'dependencies.'"[98] In brief, we can say that the function of the queen of hearts in the game of bridge is facticly correlated to the function of the queen of hearts as a card of the deck, which is, in its turn, facticly correlated to the queen of hearts as a design printed on one of the cards. What is more, there is a further relation of *Fundierung* between this design and the physical composition of the card. These examples, for Rota, demonstrate that the layering of the *Fundierung* relation is complex, since facticity is not necessarily to be identified with a physical layer. In a translation, for example, "the facticity is the facticity of words, grammar, syntax, and of meaning considered as such—it is not the facticity of anything physical."[99]

Similar investigations will find an application in Rota's philosophical reflection on mathematics as part of a far broader project, which we could describe as his

[97] Rota (1991a), p. 94.
[98] Loc. cit.
[99] Ibid., p. 111.

theory of *Fundierung*. Here, his research finds—as Husserl put it—"the entrance-gate of phenomenology,"[100] that "third way" which attempts to overcome the opposition between objectivism and psychologism. Rota's is a particular interpretation of the project of Heidegger's that pursues "a more precise characterization of the concept of *Reality* in the context of a discussion of the epistemological questions oriented by this idea which have been raised in realism and idealism" (Heidegger, 1927, p. 228). Rota's theory of *Fundierung* is not expressed in a systematic text but presents itself in his writings as a collection of ideas that delineate as many projects of development; I limit myself to indicating two of them as the principal potentialities of Rota's philosophical reflection. The first regards the possibility of formalizing the relation of *Fundierung*, in "the hope that the concept of *Fundierung* will one day enrich logic, as implication and negation have done in their time. That is, *Fundierung* is a connective which can serve as a basis for valid inferences and for the statement of necessary truths" (Rota, 1989a; see 1997a, p. 172). This idea takes on a fully programmatic character in his writings *Husserl and the Reform of Logic* (Rota, 1973a),[101] *Tre sensi del discorso in Heidegger* (Rota, 1987a, pp. 123–126),[102] and *The Concept of Mathematical Truth* (Rota, 1990a),[103] in which the relation of *Fundierung* is made the cornerstone of a project of enlargement and comprehensive restructuring of the notions of formal logic, which must not be considered as "eternal and immutable, but were invented one day for the purpose of dealing with a certain model of the world" (Rota, 1974b).[104]

Rota is convinced that "our logic is patterned exclusively upon the structure of physical objects" (Rota, 1973a),[105] as set theory demonstrates, based on the relations of "containment" and "membership."[106] To clarify this thesis he uses one of his favorite examples, namely, the marble in the box. To his mind, the aforementioned relations of "membership" and "containment" represent nothing other than the last sophisticated stage of a process that started out from this banal case, which the phenomenological tradition denounces as crude and unfit for describing mathematics and human experience in general.[107] In this regard, Rota appeals to Heidegger's analysis of *Being-in*, where Heidegger criticizes the reduction of this relationship to "the Being-present-at hand of some corporeal Thing [...] 'in' an entity which is present-at-hand."[108] It is necessary to overcome a conception of logic that reduces "Being-in" to "the kind of Being which an entity has when it is

---

[100] See Husserl (1913), p. 56.
[101] See Rota (1986a), pp. 167–173.
[102] Republished as *Three Senses of "A is B" in Heidegger*, in Rota (1997a), pp. 188–191.
[103] Republished as *The Phenomenology of Mathematical Truth*, in Rota (1997a), pp. 108–120.
[104] See Rota (1986a), p. 180. See Palombi (2010), pp. 315–316.
[105] See Rota (1986a), p. 169.
[106] See Rota (1986a), pp. 169–171, 180, 249–250, and (1997a), p. 189.
[107] See Rota (1991a), pp. 218–219.
[108] Heidegger (1927), p. 79.

'in' another one, as the water is 'in' the glass, or the garment is 'in' the cupboard,"[109] or the marble is in the box. On this basis Rota criticizes the limits of set theory, which interprets Being-in exclusively in terms of membership or of containment (Rota, 1987a).[110] To overcome them "one might begin by formalizing the ontologically primary relations of being," which were suppressed when these relations were broached (Rota, 1973a).[111] Among these "ontologically primary relations," one of the most important is precisely the relation of *Fundierung*, which Husserl's analysis "proved not to be reducible to containment or membership of sets" (Rota, 1987a, p. 34).[112] These considerations point to the second possible line of development of the theory of *Fundierung*, namely, the beginning of a course capable of shedding new light on the problem of the Nothing, particularly in its connection with that which Heidegger describes as "ontological difference."[113]

Both lines are motivated by Rota's basic conviction that the relation of *Fundierung* is the fundamental eidetic law of constitution[114] that allows us to grasp the sense in words, and to confer a sense upon the world. We might say, then, that every sense is the sense of a certain function that, out of eidetic necessity, is always founded on factic moments. In *Fundierung* Rota believes it is possible to find the worldly structure of sense.

---

[109] Loc. cit.

[110] See Rota (1997a), p. 189. See Palombi (2009b), pp. 108–110.

[111] See Rota (1986a), p. 171.

[112] See Rota (1997a), p. 189.

[113] See Rota (1987a) and (1991a), pp. 293–322.

[114] Gian-Carlo Rota, conversation with the author.

CHAPTER THREE

✧ ✧ ✧ ✧

Phenomenology and Mathematics

"Our concern will not be to discover what mathematics is made of; rather, our objective will be to discover the conditions of possibility of that eidetic domain that is called mathematics."

—Gian-Carlo Rota

The influence of scientific research on Rota's philosophical style, discussed in Section 1.2, is particularly important in his writings on the philosophy of mathematics. His particular interpretation of mathematical entities clarifies the whole of his philosophical perspective and sheds light on a number of his writings that are difficult to interpret. Later on, I shall attempt to show how, while coming from different intellectual traditions, the reflection of Imre Lakatos (1922–1974) and that of Rota meet and articulate, reaching conclusions that sustain one another reciprocally, casting an intriguing light upon the history of mathematics and upon the contribution that conventionalism furnishes us for its understanding. As with all historiographic labels, the term "conventionalism" must also be taken with due precautions. As used here, it refers principally to a constellation of concepts regarding the economy of thought and the historicity of science, and to the names Pierre Duhem, Henri Poincaré, and (partially) Ernst Mach.[1]

Up to now I have attributed great importance in this book to the critique of objectivism. Now, however, I must make it clear that mathematics is itself not exempt from this critique since, as Husserl tells us, it was mathematics that "made

---

[1] This cataloguing is accepted by a number of scholars (see, for example, Oldroyd, 1986, pp. 169–208, and Losee, 1972, pp. 143–157). It must, however, take the numerous differences between these authors into account, as certain cases in which Duhem distances himself from and criticizes Poincaré attest (see Duhem, 1906, pp. 212, 216). In particular I shall concern myself with the economic dimension, more than the dimension of social convention referred to (explicitly or implicitly) by the etymological meaning of the term.

for the first time an objective world in the true sense" (Husserl, 1959, p. 32).[2] This remark (at least in principle) is not in conflict with Rota's position that certain characteristics of our primordial experience of the world are irreducible to any form of quantification, which is nonetheless highly complex because it intersects with the exemplary character of the constitutive function of mathematics. Rota agrees with Husserl in his criticism of the "quantitative prejudice" that induces us to "feel that everything can be quantified in one way or another. Our experience with mass, distance, number, etc. leads us to transfer quantifying from the realm of factual representation to that of attempting to quantify less tangible items" (Rota, 1991a, p. 23; see also Husserl, 1959, pp. 34–41). Despite these complexities it is clear that Rota interprets phenomenology as the attempt to overcome a philosophy of mathematics impaired by objectivism and psychologism;[3] he reinterprets this Husserlian dichotomy in order to grasp the dynamics that subtend mathematics and its freedom.[4]

In his analysis of mathematics and of its relation with objectivism Rota follows the course of the most classical philosophical thought; the problems are not resolved but, rather, dissolve. The philosopher "does not attempt to dispel the mystery but, rather, observes that the mystery is 'the same' as several other mysteries" (Rota, 1987a, p. 126).[5] It is important to understand fully the sense of this process of dissolution in the framework of the classical philosophical problems that create uneasiness when we attempt "to spell out what kind of argument might be acceptable as a 'solution to a problem of philosophy.'"[6] From a certain point of view the very term "solution" has been borrowed from mathematics, and then employed in philosophy, presupposing analogies whose validity has not yet been demonstrated.[7]

---

[2] Husserl dedicates to the analysis of the role of mathematics in the creation of objective science the celebrated pages of "Galileo's mathematization of nature" (ibid., pp. 21–59).

[3] See Sokolowski (1997), p. XIII, Palombi (2009a), pp. 256–258, and Costa (2010), pp. 23–25. The expression "psychologism" is a term that originated in the 19th century that met with some success even before Husserl. Gioberti considered it to be one of the fundamental characteristics of modern philosophy (see Gioberti, 1840, Vol. II, p. 175). Husserl utilized it in the first volume of his *Logical Investigations* (Husserl, 1900–1901) to criticize radically those philosophical positions that reduce mathematics to psychological mechanisms of the knowing subject. See Rota (1990a) and (1990b); (1997a), pp. 98–99 and 117–120; and Palombi (2009a), pp. 256–258 and (2010).

[4] Hermann Weyl's perspective was similar in some respects. Weyl maintained it was necessary to overcome this dualism (Weyl, 1932, p. 77) in order to show that "mathematics is not the rigid and uninspiring schematism which the layman is so apt to see in it" but is rather a "point of intersection of limitation and freedom which is the essence of man himself" (ibid., p. 61). See also Feist (2002).

[5] See (1997a), p. 190. There is unquestionably an assonance between Rota's position and Wittgenstein's observation that "our mistake is to look for an explanation where we ought to look at what happens as a 'proto-phenomenon'" (Wittgenstein, 1953, p. 167).

[6] Rota (1990b); see (1997a), p. 101.

[7] Loc. cit.

It is evidently not a matter here of proclaiming the senselessness of the question on the basis of an analysis of language (à la Vienna Circle), but rather of grasping the fact that it has to be understood as "a universalization of the problem" (Rota, 1998f, p. 25). Investigating a question for the purpose of demonstrating the impossibility of answering it in a definite way is not an idle reflection because this oblique procedure, which does not remain focused on the obstacle but endeavors to see behind and beyond its depth, permits us to understand the nature of that specific questioning. This reflection is freely inspired by Kuhn (1962, p. 53), who wonders about the meaning of the question that has the pretense of indicating a name and an exact date for the discovery of oxygen, and by Derrida (1967a, p. 13). An oblique investigation of this kind motivates a previous study of mine as well (see Palombi, 2002). I also wish to recall here Cimatti's thesis that, by contrast, "radical questions demand radical answers, a yes or a no [...] when despite all our efforts it is not possible to give an answer of this kind, then it is perhaps opportune to change the question" (Cimatti, 2002). I believe that this position is very close to Wittgenstein (1953, pp. 21–23).

Following this style of reasoning Rota confronts us with "the double life of mathematics" (Rota, 1998f)[8] that can be translated into a more explicit question (which perhaps everyone has posed at least once in their life): are mathematical ideas invented or discovered?[9] This is a question that calls directly into play the nature of the mathematical entity, which can be interpreted as given or as constructed. This question recurs in Rota's texts, implicitly or explicitly (Rota, 1990a and 1990b),[10] for the purpose of analyzing the meaning of the question and, above all, that of its persistence (Rota, 1990b),[11] but without endorsing either of the two answers because, as in every antinomy, "the thesis, as well as the antithesis, can be shown by equally clear, evident, and irresistible proofs" (Kant, 1783, pp. 81–82).[12]

As a result, he begins to examine the question with a sidelong glance that aims first of all at understanding its conditions of possibility, describing each of the two lives of mathematics. In the first, mathematics discovers its facts as any science does, and these facts possess the same usefulness as those of other scientific disciplines and will find, sooner or later, significant applications (Rota, 1990b; see 1997a, p. 89). In this context we note one of those semantic shifts, examined in

---

[8] See Rota (1997a), p. 89.

[9] See Rota (1990b), p. 39, (1997e), p. 62, (1997a), p. 89.

[10] See Rota (1997a), pp. 108–120 and pp. 89–103.

[11] See Rota (1997a), p. 89.

[12] It is precisely the antinomic structure of the problem that leads Rota to say that "it does not matter whether mathematical items exist, and probably it makes little sense to ask the question" (Rota, 1997f; see 1997a, p. 161). In fact the problem *as such* is without consequence for mathematical research but is important for the philosophy of mathematics because it makes it possible to exercise the antinomy and indicate the inadequacy of certain traditional schools of thought. In this regard see Cellucci (2002b), pp. 137–141 and, in particular, pp. 140–141, where he quotes Rota (1997f). See also Cellucci (2009), pp. 211–213.

Section 1.2, that Rota imposes on certain terms of his vocabulary: if elsewhere the word "fact" has a negative sense,[13] in reference to a dimension completely independent of a transcendental subject, in this case it indicates a not purely subjective or psychological plane. "Fact," in this context, indicates the belonging of mathematical problems to a world that, even if without space-time existence, has to become familiar to the mathematician with a meaning not very different from the everyday one. Rota seems to pursue the ambitious goal of repeating the Heideggerian operation designed to describe a sort of Being-in-the-world of the mathematician and to repropose his intimacy, closeness, and familiarity in another context. This is one reason why he insists that "mathematical truth results from the formulation of facts that are out there in the world, facts that are independent of our whim or of the vagaries of axiomatic systems" (Rota, 1990a; see 1997a, p. 113).

It is not easy to follow Rota on the edge of phenomenological thought, suspended between objectivism and psychologism, to enter the world of mathematics. I shall attempt to grasp this delicate balance by returning to the theme from various perspectives, clarifying first what Rota saw as the "philosophical errors" that phenomenological reflection on mathematics has to avoid. He reasserts his critique of every form of "hierarchization" of reality that interprets physical objects as "more real" than those that are ideal, or that attributes a certain physical existence to every entity.[14] Rota maintains that one of Husserl's principal theses is "that of the autonomous ontological standing of distinct eidetic sciences," which consists in understanding that "physical objects [...] have the same 'degree' of reality as ideal objects (such as prices, poems, values, emotions, Riemann surfaces, subatomic particles, and so forth)" (Rota, 1973a; see 1986a, p. 169).

This proclamation seeks to avoid the symmetrical dangers as discussed in Section 2.2, inherent in objectivism and psychologism, reasserting the necessity of both poles of the *Fundierung* relation, which we have indicated as "function" and "facticity." In the specific context, this implies that mathematics and geometry are not possible without some form of notation and of writing, but that they must be neither confused with nor reduced to them. In this regard Husserl explicitly states that

> [t]he *geometer* who draws his figures on the board produces thereby factually existing lines on the factually existing board. But his experiencing of the product, qua experiencing, no more *grounds* his geometrical seeing of essences and eidetic thinking than does his physical producing. This is why it does not matter whether his experiencing is hallucination or whether, instead of actually drawing his lines and constructions, he imagines them in a world of fantasy. (Husserl, 1913, p. 16; emphasis in original)[15]

---

[13] See, in particular, Rota (1991a), pp. 2–4.

[14] See Husserl (1959).

[15] I think it is of interest to compare Husserl's reflection with that of Duhem, who maintains that for the mathematician "it matters little whether the operations he performs do or do not correspond to real or conceivable physical transformations" (Duhem, 1906, p. 20).

But we must also recall that it is precisely "the possibility of being written [*possibilité graphique*] [that] permits the ultimate freeing of ideality" (Derrida, 1962, p. 90). What, then, is the modality of access to the world of mathematics proposed by Rota? I shall tackle the first horn of the dilemma by recalling the difficulties that arise from a Platonic interpretation of mathematical entities, which Rota avoids by examining the relation that links the materiality of writing and the conventionality of axiomatic-formal representation with the plane of mathematical existence.

3.1 ✧ Mathematical Writing and *Fundierung*

The modalities of access to the world of mathematics and the distinctive characteristics of its "pure openness"[16] are connected with the theory of *Fundierung*. For Rota, mathematical theories also collect in layers that, for expediency of presentation, I suppose to be ordered linearly.[17] From this point of view any problem of mathematical research is located in one of these layers; for example, in layer *n*. Then the research belonging to *n* must take as given the results already realized in *n-1*; i.e., a researcher who studies a problem in *n* considers the results in *n-1* as ascertained facts. Thus an algebraic geometer has to take for granted the results of commutative algebra; just as in the theory of differential equations, in the real field one must be able to consider the field of real numbers as a fact. The researcher in *n* will seek to express the results of his or her research with explicit and summary formulas called "theorems," conscious of the fact that these results will be the raw material of the researchers in *n+1*. Even if a foundational research that reinterprets a branch of mathematics on new bases is possible, as Rota himself did for combinatorics,[18] a researcher cannot "run through the whole immense chain of groundings"[19] since it is potentially infinite.[20] (Note that in this case I refer to foundation in the sense of *Begründung*—grounding—and not of *Fundierung*. I shall examine later the sense in which Rota interprets the problem of foundation as ground.)

Thus, for Rota, *Fundierung* makes it possible to clarify the meaning of the layered structure that connects the various sectors of mathematics and that sustains its growth. What we have here is a problem of sense and of intentionality, and not of

---

[16] See Derrida (1962), p. 56, pp. 131–132.

[17] Gian-Carlo Rota, conversation with the author during his seminars at Strasburg University in the spring of 1996.

[18] See Rota (1995).

[19] See Husserl (1959), p. 363.

[20] "In every phenomenological regression to beginnings, the notion of an *internal* or *intrinsic* history and sense lets us delineate some safety-catches [*cranes d'arrêt*], as well as articulate, if not avoid, all '*regressus ad infinitum*'" (Derrida, 1962, p. 125). This analysis of mathematics finds its precedents in Weyl, who was influenced in his turn by the "layering of formal logic" Husserl presents in the *Logical Investigations* and in *Formal and Transcendental Logic* (see Moriconi, 1981, pp. 20–21). The reference is to Husserl (1900–1901) and (1929).

a reduction of mathematics to a single part that represents its last and lowest layer; this is not the meaning of *Fundierung* and it must not be confused with foundations.[21]

Rota examines the relation of *Fundierung* that sense entertains with verbal and written expressions and with their multiple layers (Rota, 1987a).[22] But he does not dwell on the condition of possibility of the distinctive sedimentation that characterizes mathematical knowing in which "it is of the essence of the results of each stage not only that their ideal ontic meaning in fact comes later [than that of earlier results] but that, since meaning is grounded upon meaning, the earlier meaning gives something of its validity to the latter one, indeed becomes part of it to a certain extent" (Husserl, 1959, p. 363). In his last works Husserl maintained that this condition is constituted by *writing*, which "effects a transformation of the original mode of being of the meaning-structure," permitting both its sedimentation and its reactivation.[23] The written recording of a mathematical entity is one of the conditions necessary to prevent a regression from layer to layer (this was my original premise) that would make the growth of mathematical knowledge impossible. Unfortunately Rota does not delve more deeply into this specific question, which makes it possible to cross the narrow passage between psychologism and objectivism to grasp the objective and ideal existence of mathematical entities. Phenomenology can be interpreted as an effort to "break from both conventional Platonism and historicist empiricism"[24] and thus describe "a concreter and specific history—the foundations of which are a temporal and creative subjectivity's acts based on the sensible world and the life-world as cultural world"[25] and, at the same time, insist on its ideal implications. We note that, for Derrida,

> the description of these two characteristics, ideality and spirituality, so often evoked in the *Origin [of Geometry]*, does not correspond, as we know, to any metaphysical assertion. In addition to which, they are "*founded*" in the sense of *Fundierung*.[26]

In this perspective writing founds the ideality of mathematics, but also allows it to exit the sphere of individual subjectivity; in fact, mathematical existence "does not exist as something personal within the personal sphere of consciousness; it is the existence of what is objectively there for 'everyone.'"[27] For a fuller understanding of Rota's position it is essential to keep in mind Husserl's *The Origin of Geometry*[28] and Derrida's book dedicated to it (Derrida, 1962).

---

[21] See Derrida (1962), p. 133.

[22] See (1997a), pp. 188–191; and (1997a), pp. 172–181.

[23] Ibid., p. 361.

[24] Derrida (1962), p. 59.

[25] Ibid., p. 60.

[26] Ibid., p. 61.

[27] Husserl (1959), p. 356.

[28] Ibid., pp. 353–378; Appendix VI of *The Crisis of European Sciences* is the full text of *The Origin of Geometry*. For the same text, in the same translation by David Carr, see Derrida (1962), pp. 157–180.

A further stage of Rota's reflection is his radical critique of psychologism, which, as we saw, conceives of mathematical entities on the plane of purely mental existence. In fact he is convinced that "if there is an evidence that phenomenology has conclusively hammered in, ever since the first volume of the *Logical Investigations*, it is that nothing whatsoever is 'purely psychological'" (Rota, 1973a; see 1986a, pp. 171–172). This demands the use of those "combined ways of speaking" that, as we saw in Section 1.2, Husserl judged essential to force language to comply with the exigencies of phenomenology and to attempt to think *together* an empirical and a transcendental side of the mathematical object. To understand better this existence of mathematics, objective and open to everyone, let me offer the following reflection. At a certain moment in history someone described some properties of the right-angled triangle in the form of the Pythagorean theorem. If this theorem had possessed a purely psychological reality, its "inventor" (because this would be the appropriate expression) would have been the sole repository of the theorem and he would have given a description exhaustive of every possible application and consequence. Instead, after some centuries the founders of analytic geometry developed diverse implications of the theorem, translating it into algebraic terms. Obviously these people never knew Pythagoras personally; they did, however, advance his research, which presupposes the theorem that bears his name, independently of the notations and the numerical system he used, and also independently of the fact that a man by the name of Pythagoras ever actually existed. Such considerations hold for any theorem whatsoever and constitute a basic feature of mathematics. As a matter of fact what we have here is not a mere theoretical hypothesis but, rather, a situation that has cropped up many times in history, as in the celebrated cases of Gauss.[29] Rota also recalled the more recent case of Ulam and C. J. Everett who "believed they were the first to discover the probabilistic interpretation of functional composition," completely ignoring all the previous work going back to the 19th century (Rota, 1987b; see 1997a, p. 75). Rota himself had to take this phenomenon into account when he gave a lecture on umbral calculus at Rockefeller University, during which he proposed to apply his results to the resolution of the cubic equations. The mathematician Mark Kac[xviii] recalled that at that moment he was struck by a "feeling of *déjà vu*," because he had himself discovered the method in 1930 (Kac, 1985, p. 5).

These episodes attest that the objectivity of mathematics and of science in general has, *in principle*, the character of universal transmissibility, and makes it possible to overcome all psychological individuality as well as all cultural peculiarity. The specification "in principle" indicates a non-disposable possibility that does not deny that natural or human cataclysms can affect the *de facto*, but not the *de iure*,

---

[29] Carl Friedrich Gauss (1777–1855) was the discoverer of the first non-Euclidean geometry but he avoided publishing his research for reasons of scientific advisability. Some years after János Bolyai (1802–1860) and Nikolai Ivanovich Lobachevsky (1792–1856) repeated independently the same discovery.

transmissibility of mathematics. A new barbarian invasion could destroy all the mathematics texts but not eliminate the possibility that mathematics could be discovered anew. Such a rediscovery would not permit us to speak of mathematics in the plural (many mathematics) since it (like the Pythagorean theorem) "exists only once, no matter how often or even in what language it may be expressed" (Husserl, 1959, p. 357). Hence Rota speaks of discoverers and not of inventors of theorems that constitute the facts of mathematics, which is in any event unique.

What we have said until now does not mean that different axiomatic presentations of the same mathematical object are nothing but a pleonastic and redundant aspect; rather, they refer to the adumbrations that characterize its manifestation. Just as the faces of a cube are never all directly visible at the same time, so the mathematical object always manifests itself partially. This is why, to the mathematician, "an axiom system is a window through which an item, be it a group, a topological space or the real line, can be viewed from a different angle that will reveal heretofore unsuspected possibilities" (Rota, 1988b).[30] The identity of mathematical objects is thus inseparable from the constitutive open-endedness of phenomenological research because the search for new axiomatizations presupposes the identity of the mathematical object and recognizes the fact that its properties can never be completely revealed (Rota, 1988b; see 1997a, pp. 156–157).

This thesis compels us to take up an extremely complex theme that occupied the last phase of Rota's philosophical reflection, which he described as the *primacy of identity* (Rota, 1999b; see 1997a, pp. 182–187). In his view the world is not constituted by physical objects or ideas but by all real or possible identities,[31] since such identities represent the condition of possibility of any being within-the-world. Identity is characterized by the structure of the "already" and is presupposed in all ideal or real entities, which from this point of view do not present significant differences; identity, then, is equally valid (so to speak) for stones as it is for triangles.

The "permanence of identity" is the fundamental phenomenon "upon which our relationship with the world is based," since it represents that which, in mathematical language, could be called the "undefined term."[32] These reflections confront us with some truly problematic questions that, on this occasion, I shall do no more than list. First, can that which Rota indicates with the expression "permanence of identity" be thought of as a second-degree identity? Is this sort of metarelation compatible with the primordial nature of identity? Furthermore, what is the relationship between *Fundierung*, which itself had been indicated as primitive, and identity?

We find yet another problematic implication in Rota's reference to the old "problem of existence," which has its origin in "our cravings for a physical basis of the

---

[30] See Rota (1997a), p. 156. In this regard see Section 4.2 and Cellucci (2002b), pp. 195–196.

[31] Rota (1999b), p. 112; see (1997a), pp. 185–186.

[32] Loc. cit.

world" (Rota, 1997a, pp. 185–186). It seems to me that Heidegger had already come to grips with this problem in *Being and Time*,[33] while Rota makes reference to *Ereignis*, which he reinterprets as "identity devoid of any substantial or concrete element."[34] Rota is convinced that any phenomenological investigation of the world will have to come to terms with the absence of substance that derives from the primacy of identity[35] and propagandizes this thesis with the slogan *"identity precedes existence."*[36]

Given the extreme complexity of these themes I shall do no more here than note their possible influence on the phenomenological thesis that finds in the constitution of the mathematical object the model for the constitution of every object. Before examining this problem in Section 3.4, it may be useful to examine other aspects of Rota's reflection on mathematics, comparing it with that of some scholars identified with the epistemological tradition.

3.2 ✧ The Second Life of Mathematics

The research that has hitherto been done on the factual dimension of mathematics must not make us forget that mathematics also possesses a *second life* constituted mainly by proofs (Rota, 1990b and 1997e),[37] which evidences a peculiar feature: the theorems discovered with the great intellectual commitment of generations of mathematicians are, with the passing of time, transformed into proofs just a few lines long, at the cost of further enormous efforts.[38]

The reorganization of the results of already known theories recalls, for Rota, those historico-theoretical operations that Kuhn calls "mopping-up" (Rota, 1990a).[39] This expression refers to the activity that in normal science brings the

---

[33] See Section 2.2 and Heidegger (1927), p. 247.

[34] Rota (1999b), p. 112.

[35] Ibid., p. 111.

[36] Loc. cit.; see (1997a), p. 186.

[37] See Rota (1997a), pp. 90, 140.

[38] The history of the prime number theorem represents an exemplary case of this process of simplification. See Rota (1990b); see (1997a), pp. 113–117.

The prime number theory was initially conjectured by Gauss and others, for whom, given the function $\Pi(x)$, which represents the number of primes not greater than x, the following relation holds:

$$\lim_{x \to \infty} \frac{\Pi(x)}{x/\log(x)} = 1.$$

In 1896 Jacques Hadamard (simultaneously with Charles-Jean de la Vallée-Poussin), basing his work on the theory of functions of a complex variable, succeeded in proving the prime number theorem in his 1896 paper (see Hadamard, 1896). On this subject I refer the reader to Kline (1972), pp. 829–832.

[39] See Rota (1997a), p. 116. Rota prefers the gentler expression "tidying up." The reference is to Kuhn (1962), p. 24.

facts and the predictions of a paradigm into increasingly closer agreement. It is interesting to note that, for Kuhn,

> few people who are not actually practitioners of a mature science realize how much mop-up work of this sort a paradigm leaves to be done or quite how fascinating such work can prove in the execution. (Kuhn, 1962, p. 24)

Independently of the philosophical frame of reference, this phenomenon would seem to suggest that theorems are "creations of someone's mind" (Rota 1990b; see 1997a, p. 118). It is not simple to look through this opaque process of simplification because the meaning of simplicity (as of tautology and evidence as well) is philosophically far from clear. In my remarks on this question I shall keep in mind Husserl's insistence on the fundamental importance of making mathematical evidence "into a problem" (Husserl, 1959, p. 29).

This second life of mathematics reopens the problem of giving a detailed description of the work of mathematicians that manifests both a "factual" and a "formal" aspect. This indication places Rota in a tradition, going back at least as far as Poincaré, of mathematicians and historians of mathematics who call attention to the *distinction* that must be drawn between the intention of the working mathematician, whose aim is the growth of knowledge, and that of the logician.[40] This is a crucial theoretical point that, for Rota, is decisive in the opposition between analytic and continental philosophers. In fact the *axiomatic–formal* approach, ascribable (with due caution and approximation) to the analytic school, separates the history of mathematics from the philosophy of mathematics, and becomes entangled in serious theoretical difficulties.[41] Among them we must first recall that "formalism denies the status of mathematics to most of what has been commonly understood to be mathematics." Indeed, "none of the 'creative' periods and hardly any of the 'critical' periods of mathematical theories would be admitted into the formalist heaven" (Lakatos, 1963–1976, p. 2).

Furthermore, the very *success* of Russell's logicist program, shared for the most part by the logical positivists, would have entailed—as Poincaré foresaw—"an insoluble contradiction":

> The very possibility of mathematical science seems an insoluble contradiction. If this science is only deductive in appearance, from whence is derived that perfect rigor which is challenged by none? If, on the contrary, all the propositions which it enunciates may be derived in order by the rules of formal logic, how is it that mathematics is not reduced to a gigantic tautology? The syllogism can teach us nothing essentially new, and if everything must spring from the principle of identity, then everything should be capable of being reduced to that principle. Are we then to admit that the enunciations of all the theorems with which so many

---

[40] See Lakatos (1963–1976), p. 2; and Rota (1997e), and (1997a), p. 135.

[41] For a general panorama of the aporias of the logicist program see Mangione (1976), pp. 218–230.

volumes are filled, are only indirect ways of saying that A is A? (Poincaré, 1902; see 2008, p. 31)

From Rota's standpoint we could say that the resolution of this aporia can only stem from a rigorous analysis of the factual aspect of the work of mathematicians and of their activity of complex and layered research. These reflections show the usefulness of making reference to Lakatos in order to clarify the second life of mathematics, since Lakatos—like Rota—polemically targets axiomatic reductionism. For Rota "all formalist theories [...] are reductionistic. They derive from an unwarranted identification of mathematics with the axiomatic method of presentation of mathematics" (Rota, 1990a; see 1997a, p. 112). Moreover, let us recall that Lakatos criticizes the school of mathematical philosophy which "tends to identify mathematics with its formal axiomatic abstraction" (Lakatos, 1963–1976, p. 1) and Rota takes similar issue with those who have misunderstood the role of axiomatization, "confusing mathematics with the axiomatic method" (Rota, 1990b; see 1997a, p. 96).

In fact, one of the Hungarian philosopher's most important texts, *Proofs and Refutations* (Lakatos, 1963–1976), criticizes the reduction of mathematics to axiomatics through the historical reconstruction of the research developed around Euler's conjecture[42] and the theoretical debate it provoked. Rota appreciates Lakatos's research and harshly criticizes the closed mentality of the scientific environments that were not capable of understanding its value. In particular he emphasizes that "his findings [...] were met with a great deal of anger on the part of the mathematical public who held the axiomatic method to be sacred and inviolable." As a result of this "sacrilege," Lakatos's book "became anathema among philosophers of mathematics of the positivistic school" (Rota, 1997c; see 1997a, p. 50). From this point of view the two authors find themselves on the same side of the barricade, in opposition to the so-called "formalist school" that "tends to identify mathematics with its formal axiomatic abstraction (and the philosophy of mathematics with metamathematics),"[43] and finds one of its principal philosophical points of reference in logical positivism.

---

[42] Euler's conjecture represents the generalization to a larger class of polyhedra of the relation $V - E + F = 2$ satisfied by the five regular polyhedra in which V is the number of vertices, E the number of edges, and F the number of faces. It seems that originally the conjecture had been formulated by Descartes, as shown by Descartes' own manuscript (c. 1639) "copied by Leibniz in Paris from the original in 1675–6, and rediscovered and published [...] in 1860" (Lakatos, 1963–1976, p. 6). Today we know that the conjecture "applies to 'spheroid' polyhedra, i.e., homeomorphic to the sphere, with faces homeomorphic to the disk [...]. If this condition is not satisfied, counterexamples can easily be constructed. A long exhibition of 'exceptional' polyhedra was necessary, however, to advance from Euler's original formulation [...] to Poincaré's topological version" (Motterlini, 2000, pp. 108–109).

[43] See Lakatos (1963–1976), p. 1.

Lakatos maintains that one of the clearest statements of the formalist position is to be found in Carnap, who demands that "philosophy is to be replaced by the logic of science," and "philosophy of mathematics is to be replaced by metamathematics."[44] We note that, in what is considered the manifesto of the Vienna Circle, the logical positivists boasted of their intentions to find a system of neutral formulas and a symbolism free of the "dregs" of historical languages. Their declared aim was precision and clarity, just as they rejected "impenetrable profundity." They claimed that in science there is no profundity, but only surface.[45]

By contrast, Rota and Lakatos are convinced that the concealed lemmas, the non-thematized characteristics, and the capacity to grasp "profound" and not obvious aspects play fundamental roles in history and in mathematics. As a result, the "surface" approach of logical positivism (which, for Rota, is taken up in many respects by the analytic tradition) leads to the separation of mathematics from its history, and even to the denial of this very history.[46] Hence the axiomatic-formal method does not exhaust the entire sphere of the mathematical sciences, because while it does provide an interpretation of the context of justification[47] it is inadequate for understanding the context of mathematical discovery, whose nature is eminently historical.[48] Lakatos is extremely clear in this regard, emphasizing that

> [n]obody will doubt that some problems about a mathematical theory can only be approached after it has been formalized, just as some problems about human beings (say concerning their autonomy) can only be approached after their death. But few will infer from this [...] that biological investigations are confined in consequence to the discussion of dead human beings.[49]

In this light I shall examine the importance of a particular historicity that does not coincide with the history of facts, and in which an essential role is played not only by consciousness and memory, but also by forgetfulness, obviousness, and the possibility of reactivating buried intentionality.

This viewpoint, shared by Rota and Lakatos, is useful for an analysis of mathematical practice, in which one elaborates conjectures, finds counterexamples, and, above all, chooses among a number of possible axiomatic systems. This *choice*, taken as the possibility of looking at the *same* mathematical object

---

[44] Loc. cit.; see Carnap (1937).

[45] See Carnap, Hahn, and Neurath (1929). I note that Husserl too, albeit for other reasons, criticized "profundity"; see Derrida (1962), p. 101.

[46] For the formalist, "there is no history of mathematics proper" (Lakatos, 1963–1976, p. 2).

[47] Even if it must be said that, in general, axiomatics "already supposes [...] a sedimentation of sense: i.e., axiomatics supposes a primordial evidence, a radical ground which is already past" (Derrida, 1962, p. 55).

[48] Lakatos (1963–1976), p. 2.

[49] Ibid., p. 3. For a very interesting comparison, see Foucault (1963), pp. 124–148, "Opening up a Few Corpses."

from different perspectives, is a fundamental aspect of proof, which cannot be grasped by those who champion an ahistorical conception of mathematics. Not by chance, among the formalists "as a consequence of the unhistorical conception of 'formal theory' there has been a lot of discussion as to what constitutes a respectable formal system out of the immense multitude of capriciously proposed consistent formal systems which are mostly uninteresting games."[50] For Lakatos, "formalists had to disentangle themselves from these difficulties." Their ideological choice had to be overturned, showing (as Lakatos does with Euler's conjecture) that formal systems can become so only on the basis of previously elaborated informal theories.[51]

Mathematicians do not arrive at a proof simply through the static analysis of a system of axioms defined once and for all, but rather through long and twisting routes that lead them to examine their problems on the basis of fields of mathematics that are very far from the one in which they arose. The proof of Fermat's last theorem[52] is, in Rota's view, a further confirmation of his thesis; Wiles's procedure (Wiles, 1995) is based on numerous and disparate fields of mathematics, ranging from algebraic geometry to the theory of elliptic functions utilized in the study of planetary motion (Rota, 1997e).[53]

Rota shares most of Lakatos's conclusions regarding the so-called method of "conjectures and refutations,"[54] precisely because the history of Euler's conjecture shows that a great many mathematicians, ancient and modern alike, have maintained that "conjectures (or theorems) precede proofs in the heuristic order."[55] Rota, too, takes this historic case into consideration, emphasizing how "such a formula was believed to be true long before any correct definition of a polyhedron was known" (Rota 1990a; see 1997a, p. 111). A certain preliminary knowledge is indispensable because it guides and motivates the presentation; without such knowledge we would be confronted with a paradox of knowledge similar to the one formulated by the eristics of which Plato speaks in the *Meno*:

> But, Socrates, how will you look for something when you don't in the least know what it is? How on earth are you going to set up something you don't know as the object of your search? To put it another way, even if you come right up against it, how will you know that what you have found is the thing you didn't know?[56]

---

[50] Lakatos (1978b), p. 61.

[51] Loc. cit.

[52] Fermat's last theorem states that there are no whole numbers n, x, y, z such that $x^n + y^n = z^n$, with $n > 2$.

[53] See Rota (1997a), p. 140. For an overview of Fermat's last theorem and of Wiles's proof, see Singh (1997).

[54] Lakatos (1963–1976), pp. 70–76.

[55] Ibid., p. 9.

[56] Plato, *Meno*, 80d, in the translation of W. K. C. Guthrie, London: Penguin Books, 1956.

This, then, is the reproposition and reactualization of a philosophical problem that Rota confronts by criticizing the perspective for which theorems, at least *in principle*, are indirect ways of expressing tautologies. In his view this thesis has to be reinterpreted by thinking that the "intricate succession of syllogistic inferences by which we prove a theorem, or by which we understand someone else's theorem, is only a temporary prop that *should*, sooner or later, *ultimately* let us see the conclusion as an inevitable consequence of the axioms" (Rota, 1990a).[57]

Then again, it was Poincaré himself who emphasized that no theorem would be truly new if its proof did not—*implicitly* or explicitly—involve some assumption that is considered particularly significant for the consequences that derive from it; in other words, these concealed principles are justified by the force of the theorems they make it possible to prove. By contrast, the starting conjectures are justified by the proofs that had recourse to them. In this perspective formal proofs "come almost as an afterthought, as the last bit of crowning evidence of a theory that has already been made plausible by nonformal, nonreductive, and at times even nonrational discourse."[58]

Analogous considerations also emerge in the teaching of mathematics, since "no self-respecting teacher of mathematics can afford to pawn off on a class the axioms of a theory without giving some motivation, nor can the class be expected to accept the results of the theory (the theorems) without some sort of justification other than formal verification" (Rota, 1990a; see 1997a, p. 111). In other words, we could say that logic is sufficient neither for mathematical discovery nor for the teaching of mathematics[59] if one wants to avoid reducing it to a "vain logomachy."[60]

In essence, every level of mathematical practice is involved in a "back-and-forth game" that, for Rota, represents a specific variant of the *hermeneutic circle* (Rota, 1990a).[61] Rota, here, is obviously inspired by Heidegger's reflection designed to ac-

---

[57] See Rota (1997a), p. 110.

[58] Ibid., p. 111. As far as Lakatos's heuristics is concerned, we should recall that, moving from a counterexample to the original conjecture and/or to the proof, it aims at individuating the concealed lemma that is responsible, in order to modify and, possibly, replace it; thus the hermeneutic and the critical aspects intersect. For examples relative to Euler's conjecture on polyhedra see Lakatos (1963–1976), which also provides other cases dealing with the history of mathematical analysis. An example I find quite significant is the one relative to the employment of Dirichlet's principle in the context of the theory of complex variable functions. In 1870 Weierstrass criticized the principle by means of a counterexample, for which, however, an implicit assumption within the Riemannian approach was responsible.

[59] See Poincaré (1905), pp. 216–217. When Rota writes that "an axiomatic presentation of a mathematical fact differs from the fact that is being presented as medicine differs from food" (Rota, 1990b; see 1997a, p. 96) he is quite probably thinking of Poincaré, for whom intuition has to conserve a role of "antidote of logic" (Poincaré, 1905, p. 217).

[60] Loc. cit.

[61] See Rota (1997a), p. 111; and (1993a), p. 19. I often discussed with Rota that which he described as a "pre-axiomatic grasp" (Rota, 1997f; see 1997a, p. 155). At Cortona, in Tuscany, in the summer of 1998, Rota had the chance to read a first preliminary draft of this chapter. He made some

count for the "hidden circularity in formal mathematical exposition,"[62] recovering its profound sense and defending it from the attempt to reduce it to vain tautology. Rota, then, proposes a thought experiment:

> Suppose you are given two formal presentations of the same mathematical theory. The definitions of the first presentation are the theorems of the second, and vice versa. This situation frequently occurs in mathematics. Which of the two presentations makes the theory "true"? Neither, evidently: what we have are two presentations of the *same* theory.[63]

Also in this case Rota criticizes the inadequacy of those philosophers who want to hide the dust by sweeping it under the rug, and who "instead of focusing on this strange circularity, [...] have pretended it does not exist."[64] In extreme synthesis one could say that Rota strives to graft onto the philosophy of mathematics Heidegger's lesson, which runs: "if we see this circle as a vicious one and look out for ways of avoiding it [...] then the act of understanding has been misunderstood from the ground up [...]. What is decisive is not to get out of the circle but to come into it in the right way" (Heidegger, 1927, pp. 194–195).[65] Obviously this operation is exposed to substantial criticism on two fronts: that of the Heideggerians and that of the analytics (both of strict observance) who might (for once) be in agreement in rejecting the proposal as a dangerous misunderstanding. However one wishes to judge it, this proposal represents one of the novelties in Rota's philosophical thinking, which envisions a hermeneutical interpretation of the sciences within the field of mathematics.[66]

It must be said that, above all for the peculiarities of Rota's style I have evidenced at length, this novelty is not always expressed in his texts in a linear and clear way. This limit is brought to light by the considerations made by Carlo Cellucci who, criticizing some reflections contained in Rota's essay *The Pernicious Influence of Mathematics upon Philosophy* (Rota, 1990b),[67] gives us a well balanced critique

---

notes in the margins, correcting some of his own remarks, insisting on the partiality of some of his conclusions, proposing to weaken them and render them more sophisticated (Rota, 1998f, p. 21). These notes are particularly valuable because they were made on the last occasion I had to meet him in person. I attempted to follow his indications, reproposing the theme of the pre-axiomatic grasp within the hermeneutic circle.

[62] Rota (1990b); see (1997a), p. 97.

[63] Ibid., pp. 97–98.

[64] Ibid., p. 97. Kuhn, too, defends the importance of the analysis of circular argumentation, asserting that "not all circularities are vicious" (Kuhn, 1962, pp. 176, 208), even if, in another passage, he reduces the value of such argumentation to "persuasion."

[65] A thesis close to the one I have advanced (for mathematics) is taken into consideration in Vozza (1990), p. 5.

[66] For a general overview of this problem see Ferraris (1988), pp. 311–360, and Vozza (1990), pp. 3–24.

[67] See Rota (1997a), pp. 89–103.

of analytic philosophy. Cellucci observes that the error of the analytics does not consist, as Rota seems to indicate, in a slavish imitation of mathematical reasoning, but rather in a misunderstanding of such reasoning. Hence he evidences aspects of mathematical research characterized by circularity, analogies, metaphors, and hybridizations between different fields,[68] referring to examples from the history of science, to Poincaré's thoughts on mathematical beauty,[69] and, in particular, to Lakatos's celebrated studies of Euler's conjecture.[70] These reflections clear the ground for the denunciation of the "seven deadly sins of analytic philosophy."[71]

Rota would certainly have been in agreement with what Cellucci described as "corrections" to his argumentations; it is not fortuitous that both men refer, in the same sense, to none other than Lakatos (Rota, 1997c; see 1997a, p. 50). Why, then, is Cellucci driven to make these rectifications? I believe by a need for clarity that, at times, conflicts with Rota's style. Rota, who is often allusive and fails to state his sources, produces (also profound) transformations of meaning in his quotations extrapolated from their context and from his overall discourse.[72] In fact Cellucci's observations take their cue from Rota's claim that "whereas mathematics *starts* with a definition, philosophy *ends* with a definition" (Rota, 1990b).[73] If we go no further than this formulation, then Cellucci's "corrections" are justified when he insists that this difference is inadmissible and that, rather, one has to criticize the "double blunder" of the analytics in interpreting philosophical activity erroneously, and in claiming to model it on a "false image of mathematics" (Cellucci, 2002a, p. 127).

As a matter of fact, Rota is fully aware of this problem and intends, first, to criticize the reduction of mathematics and of philosophy to their logical foundations.[74] Second, he takes up an indication of Husserl's that "one cannot define in

---

[68] Cellucci (2002a), pp. 126–127.

[69] Ibid., p. 127.

[70] Ibid., p. 126.

[71] Ibid., pp. 127–133.

[72] My working so closely with Rota often gave rise to the contagious propagation of this allusiveness, from which in fact I now suffer. On account of this style, Cellucci (rightly) wonders about the meaning of a passage in which I describe Rota's heavy attacks on analytic philosophy as a "jest" [*scherzo*] (Palombi, 1997, p. 269). As a matter of fact, the remark was an ironic allusion to a friend of mine—a mathematician and student of Rota's—who was as fascinated as he was disquieted by Rota's philosophical investigations (see Rota, 1999c). Egged on by Rota, I attempted to create an *entr'acte* (I don't know how successfully) in which his students and friends could recognize him (and themselves) in a particular crisis of "philosophical anxiety" caused by phenomenological slogans. This—Rota believed—was a symptom that could be cured not with argumentation but only with reassurance; and this is the origin of the little scene in which the philosopher Rota's follower slaps one of his mathematical followers on the shoulder, exhorting him (therapeutically) to take it easy because the master "jests" [*scherza*].

[73] See Rota (1997a), p. 97.

[74] A position advanced also in Cellucci (2002), p. XI, where he quotes Rota (1997f).

philosophy as in mathematics; any imitation of mathematical procedure in this respect is not only unfruitful but wrong, and has most injurious consequences" (Husserl, 1913, p. XXIII).

From this standpoint Cellucci's observations encourage me to emphasize the existence of significant aspects of the thinking of Poincaré, Lakatos, and Rota that converge in reevaluating the essential role that non-"superficial" (in the logical positivist sense) truths play in mathematics. I thus intend to follow their trail, in order to rethink the historicity of mathematics as alternative to the aporias in which formalism becomes entangled. However, before taking this step we have to stop and examine a classic epistemological distinction: the one between mathematical and natural sciences.[75]

3.3 ✧ Mathematical Sciences and Natural Sciences

The concept of proof for Rota and for Lakatos (albeit in different perspectives) regards the possibility of correlating and engaging with the facts of mathematics in a number of different ways: conjectures, refutations, and thought experiments, in particular. The importance of the first two modalities clearly emerges in Rota's mathematical research, and in particular in one of his best known mathematical conjectures. In 1967 he conjectured that the maximum size of an antichain in the lattice of the partitions of a finite set ought to be obtained at the level of the lattice's maximum width.[76] This conjecture was confirmed for sets constituted by a maximum of 20 elements, but in 1978 Rodney Canfield showed, by means of non-constructive methods, that for sufficiently large sets it is false. A number of articles have been written since then to determine the smallest size of a set for which Rota's conjecture is false (Rota, 1995, pp. 571–576).

Then, we can also interpret mathematical proof as "a thought-experiment [...] which suggests a decomposition of the original conjecture into subconjectures or lemmas, thus embedding it in a possibly quite distant body of knowledge" (Lakatos,

---

[75] Preliminarily, I would like to recall the fact that not all the authors I use on this theoretical journey accept my basic premise. For example, Duhem is not in agreement with the weakening of the distinction between mathematical and natural sciences proposed by Rota; in this regard see Duhem (1906), pp. 267–269.

[76] I shall limit myself to presenting a few brief definitions to clarify the meaning of this example.

- A set is said to be *partially ordered* when a partial order relation is defined on it. An intuitive example of a partially ordered set is a genealogical tree where the order relation is "being the ancestor of." A *lattice* is a particular type of partially ordered set.
- A *chain* is a subset of a partially ordered set in which no two elements are incomparable. Thus, for instance, no two (distinct) brothers can belong to the same genealogical chain.
- A subset of a partially ordered set is said to be an *antichain* if no two of its elements are comparable (with respect to the order relation). Thus, for instance, no pair (father, brother) can belong to an antichain.

For further study see D'Antona (1999b).

1963–1976, p. 9.)[77] This role of the thought experiment, whose importance Rota notes on various occasions,[78] clarifies the sense of the process of simplification because it makes it possible to establish a connection with the dynamics of the natural sciences and of technology. This, in turn, makes it possible to unify, improve, and perfect discoveries and results initially obtained in widely separated sectors of research at the cost of a great waste of energy and effort. The reevaluation in the domain of proof of the role of experiments (and not only thought experiments, as we shall see) is a further way of criticizing a rigid distinction between mathematics and natural sciences, which is often based precisely on the alleged nonexperimental nature of mathematics.

This distinction, if taken in absolute terms, does not hold in either of the two directions; in fact, if it is true that mathematical proof can be founded on experiments (thought experiments or not), experiments play a fundamental role in physics as well. Just as in physics the experiment must be analyzed in order to conjecture on the reasons for its failure and on possible remedies, so too in mathematics one must analyze that type of thought experiment which is an informal proof, in order to reveal the tacit assumptions that can give rise to a paradoxical outcome or even a contradiction.[79]

If this is how things stand, it is legitimate to treat the development of mathematics as a growth of knowledge analogous to that of the natural sciences, given the fact (as Mach insisted in 1905) that "the great apparent gap between experiment and deduction does not in fact exist";[80] and, for Rota, from this point of view "mathematical truth is philosophically no different from the truth of physics or chemistry" (Rota 1990a; see 1997a, p. 113). So, we can now return to our original question about the nature of mathematical entities and complete its dissolution, affirming, with Kuhn, that the "distinction between discovery and invention or between fact and theory will [...] immediately prove to be exceedingly artificial" (Kuhn, 1962, p. 52).[81]

This, obviously, has its price—mathematics is thus divested of the supposed aura of infallibility in which at times it is enveloped in official presentations and

---

[77] Rota is quite close to Lakatos on this when he says that "a theory meant for one type of problem is often the only way of solving problems of entirely different kinds, problems for which the theory was not intended" (Rota, 1990a; see 1997a, p. 114). Furthermore, we recall the importance Cellucci attributed to hybridization understood as "inference through which properties of objects in a certain domain are transferred to objects in another domain [...] giving rise to a partial overlapping of the two" (Cellucci, 2002b, p. 285).

[78] Rota refers to this question using at times the German expression *Gedanken-experiment* (Rota, 1985e, p. 141), at other times "thought experiment" (Rota, 1990b; see 1997a, p. 97), and in still other cases "thought experiments (what Husserl called *eidetic variations*)" (Rota, 1989a; see 1997a, p. 173; and Rota, 1998f, p. 23). For a critique of this interpretation of the thought experiment see Duhem (1906), pp. 201–202.

[79] See Mach (1905), p. 145.

[80] Loc. cit.

[81] On the same question see Kuhn (1962), p. 66.

finds itself on par, in many respects, with the other sciences. This is already evidenced by Mach's proposal to consider the principles of mathematics to be empirically derived, or by Poincaré's, to present them as free conventions. In the first case, mathematics goes through vicissitudes analogous to those of the natural sciences; in the second, certainty would be guaranteed only through a decision-making process that is substantially arbitrary. Once again, is the alternative we are presented with inescapable, or is there a theoretical route that overcomes the limitations of both? Poincaré himself attempted to consider these two apparently conflicting options together, and at the same time to distinguish his own viewpoint from that of Edouard Le Roy (1870–1954), recalling that

> [i]t is impossible to study the works of the great mathematicians, or even those of the lesser, without noticing and distinguishing two opposite tendencies, or rather two entirely different kinds of minds. The one sort are above all preoccupied with logic [...]. The other sort are guided by intuition [...]. The two sorts of minds are equally necessary for the progress of science. (Poincaré, 1905, pp. 210–212)[82]

In this regard he recalls the examples of Charles Méray and Felix Klein;[xix] the former sought to prove that an angle is always subdivisible (a truth apparently knowable by direct intuition), while the latter, in order to study one of the most abstract questions of the theory of functions (knowing whether on a given Riemann surface there is always a function that admits given singularities), studied the distribution of electric current on a metallic surface (Poincaré, 1905, p. 211).[83] This, then, was Poincaré's route that, with Rota, I shall attempt to reinterpret as an aspect of the double life of mathematics. Precaution is necessary by emphasizing, once again, the theoretical character of my reflection that can be hindered by several conflicting positions of the two authors. We recall, for example, Poincaré's thesis that there are few resemblances between the spirit of the mathematician and that of the natural scientist.

3.4 ✧ "Mathematics Is Forever"

The comparison between Lakatos and Rota developed here is not meant to suggest a sort of identification between their points of view, which, on the contrary, present significant differences. In fact, if Rota accepts the method by "conjectures and refutations," his judgment on the method by "proofs and refutations" is quite different (Lakatos, 1963–1976, pp. 47–50). He is adamant in maintaining that, "once solved, a mathematical problem is forever finished: no later event will disprove a correct solution"; "the results of mathematics are definitive [...] Mathematics is

---

[82] For his criticism of Le Roy see ibid., pp. 321–326.

[83] This analysis of the different attitudes of mathematicians with regard to proof is analogous in several respects to Duhem's distinction between broad and deep minds.

forever" (Rota, 1990b; see 1997a, pp. 93 and 101).[84] His refusal to subscribe fully to Lakatos's theses is due to his awareness of the difference between the standards of logical correctness in the 19th and the 20th centuries. In fact he maintains that Lakatos's reflections on the process of "rigorization" of mathematics[85] do not seem to appreciate fully the current standards of rigor.[86] We note that John Worral and Elie Zahar, the two epistemologists who edited the English-language edition of *Proofs and Refutations*, maintain that "the drive towards 'rigor' in mathematics was [...] a drive towards two separate goals," represented by "first, rigorously correct arguments or proofs [...] and, secondly, rigorously true axioms." Only "the first of these goals turned out to be attainable (given, of course, certain assumptions)."[87] Let me recall Poincaré's quote, so dear to Lakatos, that our fathers too, "at each stage of evolution," thought they had reached absolute rigor. "If they deceived themselves, do we not likewise cheat ourselves?"[88]

Now, let me return to Rota, who intended to dissolve the dilemma of the double life of mathematics by showing that this condition, which apparently represents one of its specific characteristics, is, indeed, an aspect that is proper to *all* science. In fact, "both mathematics and natural sciences have set themselves the same task of discovering the regularities in the world."[89] From this point of view "it matters little that the facts of mathematics might be 'ideal,' while the laws of nature are 'real',"[90] since "any law of physics, when finally ensconced in a proper mathematical setting, turns into a mathematical triviality." Rota in this regard insists that "the philosophical theory of mathematical facts is [...] not essentially distinct from the theory of any other scientific facts, except in the phenomenological details. For example, mathematical facts exhibit greater precision when compared to the facts of certain other sciences, such as biology."[91]

Thus the application of the mathematical construction of reality is based on the dual nature of reality itself, constituted by a subjective and by an objective aspect.[92] In this regard Rota maintains that the physical object must no longer be taken as a model of reality because "the techniques of phenomenological and existential

---

[84] The Italian translation/edition of this passage (Rota wrote nearly all his texts in English, but then made major changes—rethinkings—in the Italian "versions" of his works) reads (translated literally back into English): "The results of mathematics are immutable. [...] Mathematics does not betray."

[85] Lakatos (1963–1976), pp. 55–56.

[86] Gian-Carlo Rota, conversation with the author.

[87] Lakatos (1963–1976), p. 56, (editor's note).

[88] Ibid., p. 52. Lakatos quotes Poincaré (1905), p. 214. For more on the problem of the correctness of proofs see Cellucci (2002b), pp. 101–108.

[89] Rota (1990a); see (1997a), p. 118.

[90] Loc. cit.

[91] Ibid., p. 119.

[92] See Weyl (1932), p. 61.

description [...] at one end serve to bracket the physical world and thereby reveal its *contingency*, [while] at the other end they bring out the experiential reality of ideal phenomena which used to be—and still largely are—equivocally reduced to their physical shadows" (Rota, 1973a; see 1986a, pp. 169–170; emphasis in original). Rota does not limit himself to proclaiming the need for change but also attempts to propose a new model, breaking the isolation of mathematics. In fact, he is quite explicit in claiming that "the ideal of *all* science, not only of mathematics, is to do away with any kind of synthetic *a posteriori* statement and to leave only trivialities in its wake" (Rota, 1990a; see 1997a, p. 119; emphasis in original). In his view "the ideal of all science is to become mathematics," since mathematics is "the end of science" (Fontana, 1996, p. 82).[93] This effectively radicalizes the Husserlian position, indicating in the mathematical object the model of constitution of all objects. This thesis finds its motivation in one of Rota's first philosophical writings (never disclaimed), in which, analyzing the evolution of the notion of set, he maintained that

> [t]he entire phenomenological theory of object-constitution (lucidly developed by R. Sokolowski) is patterned after this typical—and admittedly oversimplified—scheme: from number to set via equivalence class. It is a central claim of genetic logic that *every* ideal object can be similarly analyzed. (Rota, 1973b; see 1986a, pp. 250–251; emphasis in original)

There is no question that the mathematical object, in many respects, is the privileged example of Husserl's reflection because it is *ideal* and, as such, "its being is thoroughly transparent and exhausted by its phenomenality."[94]From this point of view, "absolutely objective, i.e., totally rid of empirical subjectivity, it nevertheless is only what it appears to be."[95] Hence, "the question of knowing whether, for Husserl, the mathematical object is the mode of every object's constitution" and assessing the consequences of such a hypothesis[96] is a controversial matter that has been heatedly and repeatedly debated.[97] Rota's analysis (with echoes of this debate) fits disciplines such as rational mechanics perfectly, but can appear dubious for other sectors, even if he expresses hope that all the sciences will be able to "go through an independent Galilean process of concept-formation," which would give them rigor without, however, uncritically limiting physical science (Rota, 1973a; see 1986a, p. 172).

---

[93] We note, in this regard, Foucault's position, which is diametrically opposed to this interpretation of mathematics. In fact he maintains that "mathematics has certainly served as a model for most scientific discourses [...] but for the historian [...] it is a bad example, an example at least from which one *cannot* generalize" (Foucault, 1969, p. 189; emphasis added).

[94] Derrida (1962), p. 27.

[95] Loc. cit.

[96] Loc. cit., footnote 4.

[97] Loc. cit.; see Sokolowski (1964), pp. 6–36; and Biemel (1959), pp. 63–71.

Apart from its intrinsic difficulties, this thesis provides a dual indication for an understanding of the genesis of Rota's reflection, which is the fruit not only of the interaction of mathematical research with his philosophical reflection, but also of the radicalization of certain theses expressed by Sokolowski. In the last-mentioned passage Rota cites one of Sokolowski's most important texts, *The Formation of Husserl's Concept of Constitution* (Sokolowski, 1964),[98] in which Sokolowski maintains that as early as *Philosophy of Arithmetic* (Husserl, 1891) "the concept of constitution is operative in that work. It is clearly used by Husserl in the explanation of the origins of groups, one of the steps in his treatment of numbers and their sources."[99] Nevertheless in the same book he remarks that "it is correct to say this, but we cannot infer too much from it. Although numbers and categorical objects are the first stimulus to constitutional analysis, the basic model for constitution [...] is not the process of categorical activity leading to such objects."[100] It would appear, then, that Sokolowski's position is considerably less bold than Rota's.

In conclusion, I note that these reflections have great relevance for one of the original themes of this book, namely, the responsibility of mathematics in the constitution of objectivism. In fact, the weakening of the distinction between mathematical and natural sciences entails the redistribution of this responsibility over the entire body of science. (Rota, 1973a; see 1986a, p. 172). In this case, too, we find the specificity of Rota's reflection, which criticizes the analytic-positivist image of science and places in *epoché* the world of the scientific and mathematical culture to which he himself belongs. Rota's is a theoretical operation that pits itself against a precise and specific epistemological tradition because that tradition does not place the scientificity of science at a sufficiently high level. Rota, interpreting the phenomenological tradition, strives to think the very ideal of scientificity differently.

3.5 ✦ Historical Footsteps

Rota is convinced that "every mathematician will agree that an important step in solving a mathematical problem, perhaps *the* most important step, consists of analyzing other attempts, either those attempts that have been previously carried out or attempts that he imagines might have been carried out, with a view to discovering how such 'previous' approaches failed" (Rota, 1990b; see 1997a, p. 99). This important statement motivates his frequent concerns about the situation of the history of mathematics, which he termed "an enticing but neglected field" (Rota, 1974c).[101] This situation depends not only on the current political and cultural context but also on the very nature of mathematics, which, like other sciences, "does not admit

---

[98] See in particular the first chapter (pp. 6–36), "Constitution and Origins of Numbers."

[99] Sokolowski (1964), p. 35.

[100] Ibid., p. 202.

[101] See Rota (1986a), p. 157.

a history in the same sense as philosophy or literature do. An obsolete piece of mathematics is dead to all but the collector of relics."[102]

But in what sense, then, can we speak of a history of mathematics? Rota maintains that "no mathematician will ever dream of attacking a substantial mathematical problem without first becoming acquainted with the *history* of the problem, be it the real history or an ideal history reconstructed by the gifted mathematician" (Rota, 1990b).[103] In this sense "the solution of a mathematical problem goes hand in hand with the discovery of the inadequacy of previous attempts, with the enthusiasm that sees through and gradually does away with the layers of irrelevancies which formerly clouded the real nature of the problem."[104] This is why Rota insists that mathematics confirms the historical nature of thought—a position held by the "best philosophers of our century," from Husserl to Croce.[105] This ideal history "calls for the reconstruction of an intentional, not a real history, a reconstruction which can only be carried out *in the light of* a problem at hand which acts as motivation" (Rota, 1973b; see 1986a, p. 250). Such a view is founded on the possibility of reactivating intentionalities on the basis of the theorem as it is constituted in present mathematics, and in the present problem of the mathematician, in the form of "repeating an origin."[106]

In this sense comprehending mathematics and its historicity means reproposing Husserl's requirement of a "methodical production, proceeding from the present and carried out as research in the present" (Husserl, 1959, p. 371). Paraphrasing Husserl's text we can say that only on the basis of present-day needs can the mathematician understand the development of a theorem, but without an understanding of the beginnings of this development we cannot grasp the theorem's present meaning.[107] To grasp the sense of the ideal history proposed by Rota we have "no other choice than to proceed forward and backward in a zigzag pattern," as Husserl insists in *The Crisis of European Sciences.*[108]

In fact, Rota sees in Husserl's "genetic phenomenology" the instrument suitable for reconstructing an ideal history based on the assumption that

> anything that has been made into an object (in mathematics, we would say "defined") eo ipso begins to conceal the original drama that led to its constitution. Most likely, this drama followed a tortuous historical path, through things remembered and things forgotten, through cataclysms and reconstructions, pitfalls and lofty intuitions, before terminating with its objective offspring, which will thereupon naively believe itself to be alien to its origin. The reconstruction of this

---

[102] Loc. cit.

[103] See Rota (1997a), p. 99. See Lakatos (1963–1976), pp. 10–11.

[104] Rota, loc. cit.

[105] Ibid., p. 100. For a comprehensive picture of the debate on this theme see Bonicalzi (1982), pp. 11–50.

[106] See Derrida (1962), p. 34.

[107] See Husserl (1959), p. 58.

[108] Loc. cit.

genetic drama is a *logical process for which classical logic is totally inadequate.* (Rota, 1973a; see 1986a, p. 170; emphasis in original)

Here again, apropos of ideal history, Rota appears to see eye-to-eye with Lakatos, who justifies the dialogical form of his *Proofs and Refutations* with the intent to "contain a sort of *rationally reconstructed or 'distilled' history*" (Lakatos, 1963–1976, p. 5; emphasis in original). Lakatos takes as a point of reference Poincaré who, in his turn, looks to the biological law of Ernst Haeckel (1834–1919) to understand the development of mathematics (Poincaré, 1908, pp. 436–438). Nevertheless, the agreement between Rota and Lakatos holds only to a certain extent, since Rota's ideal history is never a whole with nothing left over, a development in which every phase is summarized and subsumed.[109] On the contrary, it contemplates, of necessity, erasures, "things forgotten," and periodic lapses that transform notations and proofs into relics. I shall attempt to reinterpret these reflections on the historicity of mathematics analyzed by Rota in the light of an erasure that, in a triple movement, also conceals itself.

Getting back to our analysis of the mathematical object we return to the question we posed earlier: why does science aspire progressively to transform its own knowledge into tautologies and analytic trivialities? The doubt I intend to raise regards the possibility of a correlation between this process of trivialization and the process of erasure that characterizes the historicity of mathematics. The condition of possibility of mathematics is represented by its capacity to accumulate a knowledge that does not consist in the recording of every single stage but, rather, that preserves mathematical objects by means of a powerful simplification. This simplification is the transcendental condition that emerges precisely from the apparent opposition between the *quid facti* and the *quid iuris* of the activity of mathematicians as the rhythm of an erasure, of a removal that sustains mathematics and constitutes its assumption.

Of these erasures Rota limits himself to individuating explicitly the one that tends to conceive of theorems as tautologies without, however, clarifying their conditions of possibility. Personally, I believe that such erasures can be individuated in the necessity of mathematics to free itself from the dross and dead weight that would rapidly tend to block its growth. Recapitulating the considerations I have developed to this point, I would like to investigate Rota's reflection further by thinking of mathematics as a field that develops while leaving traces that are erased three times over:

1. First, all the work through which one succeeds in completely writing a theorem is erased. As I evidenced earlier, a proof is nothing but a "fair

---

[109] In this context, it could be interesting to reconstruct the affinities and differences between Lakatos and Rota with reference, on the one hand, to the Hegelian model they share and, on the other, to the respective influences of Popper and Husserl.

copy," a final product safe from counterexamples and refutations, from which one has removed all the errors, the failed attempts, and the rejected conjectures, which, however, are essentially what made the proof possible in the first place.

2. The process of "simplification" that the discovery of every new theorem triggers does not entirely remove the previous proofs that, in the light of further motivations, can be reactivated. Nevertheless, we must recall the outcome of the lapse that over the long term transforms mathematical texts into relics that lose their interest for militant mathematicians. In this way the mathematical objects remain, but the entire chain of reaxiomatizations tends to become concealed beneath the proof that the present-day historian considers simple.

3. Finally, the logical reduction of a theorem to tautology is a further form of removal that tends to conceal the very existence of all this complex work of writing and erasure that in principle "ought not to exist."

I believe that this complex dynamic of erasure is not fortuitous, but possesses a phenomenological necessity. In fact, how can mathematics "maintain its original meaningfulness through living reactivatability if its cognitive thinking is supposed to produce something new without being able to reactivate the previous levels of knowledge back to the first?" (Husserl, 1959, p. 363). This reflection is analogous to the one I developed earlier regarding the layers of mathematics correlated by relations of *Fundierung*. If the transmission of mathematics consisted in the reproduction of every detail of the work necessary for the discovery of a theorem, or of the first proofs of that theorem (necessarily longer and more complex than the ones that will be elaborated subsequently), then the process of the transmission of mathematical knowledge would jam up within a few generations, and mathematics as an infinite field would thus be impossible. Hence the traces I have indicated are of a particular type, because they belong to mathematical concepts that are

> formed by a pattern of identification which must always erase its footsteps; as time and history proceed, layers of usage, history, tradition and self-interest are heaped on, until the concept acquires a giddy air of always-having-been-around-ness. (Rota, 1973b; see 1986a, p. 249)[110]

The law of the formation of mathematical concepts manifests itself, then, as trace that sutures oblivion (as past to be forgotten) and remembrance (intentionality

---

[110] For Heidegger there are "coverings-up which are accidental; there are some which are necessary, grounded in what the thing discovered consists in" (Heidegger, 1927, p. 60).

to be reactivated). Clearly, it is not possible to define *a priori* what part of mathematics will be subject to the process of erasure and what, by contrast, will continue to be topical in a certain epoch. In fact a proof that one historical phase finds muddled can, in another phase, be considered interesting, because it is judged from the viewpoint of its capacity to connect heterogeneous fields of mathematics. All these aspects manifest themselves in a typical fashion in the historical phases described as phases of crisis of foundations.

3.6 ✧ Epoché and the Crisis of Foundations

The word "crisis" is ambiguous, and for this reason I wish to make some lexical and philosophical clarifications before going on to my analysis of Rota's argumentations. Its meaning, both in common speech and in specialized psychological, sociological, and economic vocabulary, implies negative situations—conditions of instability or danger, or dramatic emotional upheaval. By contrast, in the medical domain the word has a "neutral" value, referring to the point in the course of a serious disease at which a decisive change occurs, leading either to recovery or to death. It is in the latter sense that Poincaré refers to the crisis of mathematical physics (Poincaré, 1905, p. 297).

To understand the positive role of a crisis we can make reference to the word's etymology, which derives from the Greek verb *krínein*, meaning to decide, separate, or judge. If etymologically "crisis" refers to choice and decision, philosophically it is also connected with the necessity of suspending and overcoming ancient convictions and beliefs; it is in some way influenced by the ancient reflection on the value of human knowledge. It has been the progress within mathematics and the logic-formal sciences—the birth and development of infinitesimal calculus, the discovery of non-Euclidean geometries, the generalization of algebra, the articulation of the theory of real variable and complex functions—that has sparked "revolutions of ideas" so stunning that philosophers of mathematics have put "crisis" at the top of their agenda. I believe that the theme of crisis does not represent the skeptic's surrender to ataraxy, but rather a way of appreciating the characteristic nature of mathematical and, more generally, scientific knowledge.[111]

In this sense the crisis of a model does not mean a general setback for science but, rather, the possibility of problematizing sclerotized conceptions, of overcoming obvious attitudes not only by conquering new theories but also, at times, by recovering and reactualizing perspectives held to be obsolete. In this sense the positive and, indeed, vital role of crisis for mathematics can, once again, be fully recovered only through the recovery of its historical dimension. This is why Rota maintains that "thanks to the present crisis of foundations, we are allowed a rare opportunity to observe fundamental scientific concepts in the detached state tech-

---

[111] See, in particular, Lakatos (1978b), p. 8.

nically known as 'bracketing'" (Rota, 1973a; see 1986a, p. 168).[112] He insists that "bracketing"—*epoché*—is the fundamental component of genetic logic that "in the exact sciences [...] is primarily the critique of foundations. Here, a proliferation of constructs, spurred on by unbelievable success—at least in the past—has led to a Babel of layered theories. Progress is made difficult by the weight of an awe-inspiring tradition" (Rota, 1973b; see 1986a, p. 249). We have repeatedly noted that Rota rereads Heidegger through the lens of Husserl. Here we have yet another example, where genetic logic is drawn near to Heidegger through the theme of tradition. Rota recalls Heidegger's statement that

> [w]hen tradition thus becomes master, it does so in such a way that what it "transmits" is made so inaccessible, proximally and for the most part, that it rather becomes concealed. Tradition takes what has come down to us and delivers it over to self-evidence; it blocks our access to those primordial "sources" from which the categories and concepts handed down to us have been in part quite generously drawn. Indeed it makes us forget that they have had such an origin, and makes us suppose that the necessity of going back to these sources is something which we need not even understand. (Heidegger, 1927, p. 43)[113]

For Rota, "uncovering the all-but-irretrievable origin is the purpose of genetic logic;"[114] but to do so one must first grasp the sense of this "loss." It is not a matter of recovering a first concept in a chronology, in an event determined by history, or in a precise instant of time; rather, it is a matter of a "loss" both primordial and constitutive that we can interpret in the light of the triple erasure seen earlier. Rota individuates a succession of phases that mark the rhythm of the historical evolution of mathematics and of science in general according to a "recurring pattern": the ideal object determines the constitution of a science that subsequently goes through a phase of crisis, in which genetic analysis that constitutes a new ideal object intervenes.[115] It is not a question, here, of a reproposition of the idea of historical recurrence from the viewpoint of mathematical historicity, or of an updated version of a philosophy of history that anticipates the succession of critical and organic periods; rather, it is a matter of striving to evidence the function of genetic phenomenology "at two stages in the development of a science: at dawn, by circumscribing an autonomous eidetic domain with its internal laws; at dusk, by the criticism of that very autonomy, leading to an enlargement of the eidetic domain."[116] Obviously this description is not to be understood in the sense of a

---

[112] Rota always preferred the term "bracketing" to express what Husserl called *epoché*, or phenomenological reduction.

[113] This quote from Heidegger is contained in Rota (1973b); see (1986a), p. 249.

[114] Rota, loc. cit.

[115] See Rota (1973a); see (1986a), p. 173.

[116] Ibid., pp. 172–173.

consciously exercised philosophical activity since Rota, while considering scientific work as a specific case of this phenomenological attitude, emphasizes that it remains for the most part unconscious and irreflexive. In fact it is not part of the scientist's task to "bracket" his practice, since he is concerned with objectifying, not with problematizing what has been objectified.[117] In any event, phases of crisis continue to be of crucial importance for the development of science since, as Heidegger also notes, the "real progress" of scientific research "comes not so much from collecting results and storing them away in 'manuals' as from inquiring into the ways in which each particular area is basically constituted."[118] In this sense "the real 'movement' of the sciences takes place where their basic concepts undergo a more or less radical revision which is transparent to itself. The level which a science has reached is determined by how far it is *capable* of a crisis in its basic concepts."[119]

In his rereading, Rota calls into question the relation between the object of mathematics and the axiomatic method that, at its extreme points, tends to confuse and, even identify the object with the method.[120] This suggests a further reference to Heidegger, who with regard to the crisis in the foundations of mathematics states that "in the controversy between the formalists and the intuitionists, the issue is one of obtaining and securing the primary way of access to what are supposedly the objects of this science" (Heidegger, 1927, pp. 29–30).

---

[117] Ibid., p. 170.

[118] Heidegger (1927), p. 29.

[119] Loc. cit.; emphasis in original. Heidegger goes on to say that "in such immanent crises the very relationship between positively investigative inquiry and those things themselves that are under investigation comes to a point where it begins to totter" (loc. cit.).

[120] In his continual work of comparison, analogy, and distinction between mathematical and natural sciences, Rota takes up the problem of foundations from the viewpoint of science in general. This problem, in several respects, is inspired by Heidegger's *Being and Time*, which contains some extremely interesting remarks regarding the theory of relativity. For Heidegger, physics, "as a theory of the conditions under which we have access to Nature itself, seeks to preserve the changelessness of the laws of motion by ascertaining all relativities, and thus comes up against the question of the structure of its own given area of study—the problem of matter" (ibid., p. 30). The problem of foundations also manifests itself in biology, where "there is an awakening tendency to inquire beyond the definitions which mechanism and vitalism have given for 'life' and 'organism,' and to define anew the kind of Being which belongs to the living as such" (loc. cit.).

Rota takes up these themes in his views on Husserl's genetic logic, interpreted as "the logic of concept-formation, designed to counter the trend towards uncritical adoption of notions of everyday life as structural concepts. In the life sciences, for example, concepts of current use such as 'life,' 'evolution,' or 'organic,' resemble, if only by their plebian origins, the four Aristotelian elements. But chemistry was born only when four well-established notions out of everyday experience [...] were replaced by artificial concepts which were not at all to be found in experience, but were instead the result of an authentic genetic search, quite different from the naive recording of natural phenomena as they appear to the senses" (Rota, 1973b; see 1986a, p. 249). Analogous considerations were developed by Bachelard (even if from a different perspective); see Bachelard (1938).

3.7 ✧ Economy

The layering of history tends to preserve the "ideal" (Rota would say "functional") dimension of mathematical entities over against their factic counterpart in the domain of a *Fundierung* relation. In Rota's words, "The relation between the truth of mathematics and the axiomatic truth which is indispensable in the presentation of mathematics is a relation of *Fundierung*" (Rota, 1990a; see 1997a, p. 112). Returning to our previous example, what is still alive today of Pythagoras's theorem is its sense determined by the correlations with all the other (potentially infinite) objects of mathematics, independent of the type of notation and the number system Pythagoras used to prove it.

It is evident that one of the infinite possible modes of facticity is indispensable, since mathematics has need of at least one type of writing, of axiomatization, of notation. Nevertheless, facticity must be seen as a simple opening of mathematics towards its possibilities, and not in its determinate specificity. One could say, paraphrasing Rota's example regarding the *Fundierung* relation, that the sense of a theorem cannot be without a certain notation, but neither can it be reduced to that notation.[121] In order to examine the characteristics of this ideal dimension more thoroughly, I have to take up a question that I endlessly debated with Rota, without ever reaching an agreement—the question of the economic dimension of science, and of mathematics in particular.

Personally, I believe that the hallmark of the law of the formation of mathematical concepts, which recapitulates also the sense of the triple erasure we saw earlier, must be sought in *economy*, not as a pragmatic but, rather, as a transcendental dimension of mathematics. In fact, the sedimentation of concepts is, essentially, an economic operation,[122] where "a behavior, an operation, an instrument" are determined in the "limits of a certain use [...]; their rules [...] come to be established." On the whole, it is a matter of a "baggage of experience [...] transmitted to the successive generation, *with which a fundamental associative connection is thereby established*" (Sini, 1965, p. 200; emphasis in original). I believe that on this (and within this) layering the process of erasure acts with an economic function that is indispensable first of all because (as we have seen) a total reactivation of all the layers of sedimentation would paralyze the life and the development of science "just as surely as would the radical impossibility of all reactivation" (Derrida, 1962, p. 105).

To examine this economic character in detail I have to begin with an analysis of some general aspects of science that are subsumed in the reflection on method as a distinctive aspect of its practice. The word "method" derives from the Greek *methodos* and from the Latin *methodus*, whose meaning refers to research, enquiry, investigation, and their modalities. The etymology of the Greek word shows that

---

[121] See Rota (1987a); see (1997a), pp. 188–191.

[122] In this regard see Sini (1965), p. 200, and Paci (1965), Sections 45–49.

it is composed of *meta*, to pursue, and *odos*, "way," and thus, literally, to follow a "way" in order to reach a certain place or goal. Hence, at first blush, "method" indicates the way, the path, the procedure followed in pursuing a goal, in performing purposeful activity, according to a predetermined order and plan in light of the end one intends to obtain. In the metaphor of traveling a road is implicit the meaning of a route that, in principle, is public and thus repeatable.

There is, however, another sense of "method" that must not be overlooked, since it refers to that which the Latin defines as *iter transversum* and the Italian as *tragitto*, meaning the shortest way or route. In Latin this significance was originally applied to the military domain to indicate the route followed by the commanders in order to confer with the sentries; subsequently, it was applied to the field of the sciences and arts. This reference to a sort of "shortcut" indicates an "economicity" of action and of route but also renders its public character less sure, or at least accepts it only in principle. In fact, anyone can take a shortcut but only if one knows *where it is*, just as the proof of a theorem is reduced to a triviality *only a posteriori*.

Although it is true that "as soon as a method is conceived, it becomes accessible to anyone," there is no sort of metamethod "to invent methods or to 'find' or create new scientific theories of interest" (Radnitzky, 1979, p. 205). This brings us back to the beginning. A method is not fortuitous; rather, it is the fruit of a procedure or of an attitude, but the choice among a multiplicity of methods is *extramethodical*. In essence, one can say that shortcuts to avoid twisting lanes and blind alleys exist only *a posteriori*, after time has already been spent and effort made. The meaning of the methodical aspect of scientificity can therefore not be that of an algorithm or a recipe; it is more profound and refers to the historical character of science (upon which Rota, too, insists) and to its economic character, of which I am so fond. The shortest or simplest way is sought (by the simplifiers during the "mopping up" process) precisely to avoid jamming the mechanism of the transmission of knowledge, but it must also take insightfulness into account. "You solve a problem because you know that by solving the problem you may be led to see new ideas that will be of independent interest. A mathematical proof should not only be correct, but insightful" (Rota, 1985a, p. 99).

At this point my reflection diverges from Rota's because to interpret the question of "economy" I shall turn to Duhem, who interprets scientific theory "as an economy of thought" and as an "enormous relief to the human mind, which might not be able without such an artifice to store up the new wealth it acquires daily." These indications of the French epistemologist's are supported by an explicit quote from Ernst Mach, who maintained that "intellectual economy" represents "the goal and directing principle of science" (Duhem, 1906, p. 21). This explains why—paraphrasing Mach—the mathematician who pursues his studies without a clear idea of the economic tendency of mathematics at times has the unpleasant sensation that paper and pencil are more intelligent than he is. Such uneasiness is in some respects similar to that of the philosopher who intends to reduce all mathematics to axiomatics and then finds himself with nothing but tautologies that, having lost their historical and intentional meaning, seem empty.

Although Mach refers in particular to experimental knowledge, he indicates that among the most important instruments for the accumulation of knowledge are language and writing; and, above all, mathematics that, in its aspects both of theory and of calculation, allows us to use the knowledge generated in centuries of research.[123] From this standpoint there are interesting points of contact with the reflection on writing in Husserl's late works, subsequently investigated by Derrida, who takes up the problem of the "word-hoarding" (*thésaurisation*) without which science and history—but above all the idea of phenomenology as an "infinite task"—would be unimaginable (Derrida, 1962, p. 78).[124]

Rota is not in agreement with the radicalization of phenomenology that Derrida proposed through the analysis of writing, but above all he rejects the references to Duhem and to Mach. Rota does nothing other than repropose the traditional Husserlian theses expressed in *Logical Investigations* that interpret the economic perspective in biologist, psychologist, and teleological terms. By contrast, I believe that the historical knowledge that also founds science possesses a structure of an economic type, because through a process of idealization I can go "from the instruments to the operations with which the corresponding intentionalities of meaning are 'presentified.'"[125] Such idealizations can be interpreted in general as "abbreviated, simplified structures [...] in which the essential meaning of certain operations is enclosed."[126]

As regards the aspect of insightfulness, we need to go back to the example of Fermat's last theorem. Rota insists that the theorem, as such, is not "of any interest whatsoever" (Rota, 1997e),[127] raising a question about the sense of the efforts many renowned mathematicians lavished for centuries. For Rota, behind this question lies an error, which consists of

> assuming that a mathematical proof, say the proof of Fermat's last theorem, has been devised for the explicit purpose of proving what it purports to prove. We repeat, appearances are deceptive. The actual value of what Wiles and his collaborators did is far greater than the proof of Fermat's conjecture. The point of

---

[123] See Mach (1883).

[124] Regarding the polemic and the subsequent clarification between Mach and Husserl on the theme of the economic dimension of science, see Mach (1883) and Husserl (1900–1901).

[125] Sini (1965), p. 201.

[126] Loc. cit.

[127] See (1997a), p. 140. Rota emphasizes that "were it not for the fact that Fermat's conjecture stood unsolved amid a great variety of other Diophantine equations which could be solved using standard techniques, no one would have paid a penny to remember what Fermat wrote in the margin of a book" (loc. cit.). The formulation of Fermat's last theorem was made by the great French mathematician in the margin of the Greek-Latin edition of Diophantus's *Arithmetica*. See Penrose (1989), p. 76.

the proof of Fermat's conjecture is to open up new *possibilities* for mathematics. [...] Future mathematicians will discover new applications, they will solve other problems, even problems of great practical interest, by exploring Wiles's proof and techniques. To sum up, the value of Wiles's proof lies not in what it proves, but in what it has *opened up*, in what it will make possible. (Rota, 1997e; see 1997a, pp. 143–144)[128]

This is why such a conjecture was able to stimulate the creation of so many objects vital for the development of mathematics. In particular we note that its proof required "a large number of intermediate steps. A great many of these steps are number theoretic lemmas which are given substantial proofs of their own. Some such lemmas are currently in the process of being granted the status of genuine new theorems of independent interest" (Rota, 1997e; see 1997a, p. 143).

Returning to the metaphor of the way and of the shortcut innate in the word "method," allow me a further metaphor: let us describe human knowledge as a territory that is connected by a network of roads. These roads can be evaluated on the basis of two criteria: the distance and the number of places they connect. Even if we can imagine significant counterexamples, by and large we can say that these two criteria are opposed. Based on the first criterion, we can imagine a man who has to reach the same destination every day, and who one day finds a shorter route than the usual one; from then on he will change his itinerary to save time and effort. This is a *conscious* saving, since it conserves the memory of having *shortened* the journey. In this example, given by Mach, the memory is not of a particular labor but, rather, of a *liberation* of labor that can be employed in a better way. We are confronted with an analogous situation when we use the concepts of science.[129] Indeed, this operation of simplification possesses a double efficacy because, on the one hand, it makes it possible to economize on energy and, on the other, it conserves the memory of this operation as well as the existence of longer and more tortuous routes that once led to the same destination.

Nevertheless, sooner or later, the faint recollection of the simple existence of these longer routes is no longer sufficient. At this point there is a need for that work

---

[128] Kuhn remarks that an analogous perspective "provides a clue to the fascination of the normal research problem. Though its outcome can be anticipated often in detail so great that what remains to be known is itself uninteresting, the way to achieve that outcome remains very much in doubt. Bringing a normal research problem to a conclusion is achieving the anticipated in a new way, and it requires the solution of all sorts of complex instrumental, conceptual, and mathematical puzzles" (Kuhn, 1962, p. 36).

[129] See Mach (1883). Mach reconstructs a brief philosophical genealogy of the concept of simplicity and of economy applied to science, individuating in Copernicus, Galileo, and Newton, its first theorizers. Clifford, Kirchhoff, and Avenarius are among the scholars who have been occupied with this theme in the past two centuries. Mach also discusses the criticism Husserl (1900–1901) made of his analysis of science as economy of thought, responding that if its application to the origin of scientific knowledge in general (what Mach called his general theory of theory) is difficult, such an analysis applied to mathematics might have greater possibility of success.

of historical-cum-intentional reconstruction that at least partially brings back the old itineraries and, above all, their stages that are significant in the light of further proofs and motivations. This intentional reconstruction often proves difficult because "mathematicians more than others tend to eliminate all trace of development as soon as they present their findings" (Mach, 1905, pp. 182–183).[130] Still, it needs to be reaffirmed that "the perfectly clear recognition of mathematical propositions is by no means attained all at once, but is preceded and prepared by incidental observations, surmises, thought-experiments, and physical experiments with counters and geometrical constructions" (Mach, 1905, p. 183).

---

[130] Personally I find that there are significant assonances between Mach and Rota here—which Rota, however, always denied, perhaps on the basis of Husserlian indications.

CHAPTER FOUR

✧ ✧ ✧ ✧

The Value of Mathematics

"Unfortunately, the reigning mediocrity of our days demands a heavy toll of any scientist who dares stray from routine publication of technicalities, over into the blue yonder of speculation."

—Gian-Carlo Rota

In one of his last writings, Gian-Carlo Rota sounds a dramatic alarm for the survival of mathematics. It was the final occasion for continuing his reflection on the value of this discipline, in keeping with a tradition that has vaunted some of the leading mathematicians of the 20th century. This final chapter, inspired by the title of a celebrated text by Poincaré, is an attempt to find in the very perils that threaten mathematics occasions to rethink its value for scientific and philosophic research. In this way we shall have an occasion to return to and investigate more thoroughly some aspects of the phenomenological sense of intentionality that were examined in Section 2.2.

Rota maintains that the queen of sciences is threatened by "ignorance of its results" and "widespread hostility towards its practitioners."[1] Among the numerous and heterogeneous causes that foster such animosity he singles out the conviction that mathematics must possess many applications.[2] This contention has been wrongly taken by a heterogeneous throng of major social figures (in politics and in the military, from entrepreneurs to educators) to mean immediate and direct applicability. But such applicability is the exception rather than the rule—a fact that generates frustration and a misguided sense of uselessness that (directly or indirectly) leads to a reduction of research funding, a fall in the number of university mathematics students[3] and, by a sort of chain reaction, fosters sterile ethical and epistemological debates on pure science and technology.

---

[1] Rota (2000), p. VII.

[2] Ibid., pp. VIII–IX.

[3] See the essay *Ten Lessons for the Survival of a Mathematics Department* in Rota (1997a), pp. 204–208.

This desire for ensured and instantaneous (we could even say "preventive") applicability is the result of numerous factors rooted in the current cultural situation, perhaps the most pernicious of them being a widespread economism, hypnotized by the present and the immediate future, which, aiming at short-term gain, ignores the history of mathematics and the history of science in general.[4]

Such a viewpoint is so narrow that "company-mindedness" would be a better term for it. Not by chance, the global and long-term vision of political economy is now being increasingly mortified by that "business economics" whose success can be measured by the spread of its aggressive terminology and of its criteria of evaluation, which judge other disciplines on the basis of the short-sighted schema of the double entry.

These criticisms do not mean rejecting any sort of planning dimension, organization of resources, and hierarchy of priorities, on the basis of an anarchist theory of knowledge taken to the extreme.[5] They intend, rather, to denounce the incapacity to range over a broader horizon that couples the short-term with the long; the incapacity to view the great technological conquests as a composition of many small heterogeneous results, apparently devoid of "applicability."

For those who judge such considerations to be obvious, I recall Rota's warning that "mathematics risks becoming a curiosity that we will soon be showing our children at the zoo of endangered intellectual species" (Rota, 2000, p. VII). It is my intention, then, to develop his sorrowful exhortations into reflections on the relation between the teleological and nonteleological aspects, and the theoretical results and technological applications, of mathematical research. My reflection avails itself of Rota's fundamental philosophical reference, represented by phenomenology, and begins with his radical critique of the myth of progress.

4.1 ✧ The "Myth of Progress"

A look at Rota's views on the concept of progress will help us bring the general framework for his critique of the "company-minded" interpretation of mathematics into focus. Rota, in fact, considered progress to be incompatible with contemporary mathematics devoid of unitariness, characterized by historical discontinuity and looking, at times, towards the past (see Rota, 1990a; see 1997a, p. 117).

This was a recurrent theme in Rota's conversations with friends, students, and colleagues, even if it was barely mentioned in his writings; it sheds considerable light on the axiological correlates of his philosophical reflections, centered on the theory of knowledge. In this regard a short section of the text of his course of

---

[4] We recall that Rota attenuates the distinction between mathematical and natural sciences; see Rota (1993a), p. 28.

[5] See Feyerabend (1975), and Palombi (2004), pp. 34–35.

lectures on phenomenology at MIT is particularly significant, in which the problem is dealt with in specifically economic terms.[6]

Progress, here, is defined as the myth that the human condition is characterized by a constant tension towards material and spiritual improvement. Rota denounces the irrationality of this conviction on the basis of the precise limits of the food, energy, and environmental resources of the Earth. Furthermore, he stigmatizes the assessment of development based exclusively on the increase of such factors as the statistical average of life expectancy, caloric intake, average income, or Gross Domestic Product. This, for Rota, is a way of thinking that considers just one part of the planet and completely neglects other fundamental aspects of human life.[7] The myth of progress would be untenable if different factors were considered, such as the increase in mental illnesses,[8] or free time, or the time parents in the West dedicate to their children. This untenability would be no less evident if we were to consider the same quantitative factors, but referred to the countries of the third and fourth world.[9]

These problems have been staring us in the face for decades, and yet one continues to regard them as momentary difficulties that will be resolved by the very progress that has created them, without structurally modifying our mode of producing, consuming, living, and thinking. Rota attributes this conviction (which in many respects is irrational) to reductionism, in the sense of an "intense desire to be able to understand by oversimplifying and reducing the world to one fundamental level" (Rota, 1991a, p. 30).

The analysis of the concept of progress and of the philosophies of history connected with it is a major theoretical and axiological question that represented a point of reference for Rota that I cannot go into here.[10] Rather, I intend to show that Rota analyzed the semantic area of the word "progress" in order to underline the teleological components. In this perspective the concept of "progress" becomes a philosohpical interpretation of human life that flattens every aspect of existence on a vector drawn towards an end.

4.2 ✧ Intentionality Is Not a Teleology

Rota opposes this cumbersome primacy of teleology in the interpretation of the growth of scientific knowledge not only on the terrain of history, economics, and

---

[6] Rota (1991a), pp. 29–30.

[7] Ibid., p. 30.

[8] The question was already raised by Freud; see (1929–1930).

[9] For chronological reasons, and for reasons of academic and cultural contiguity, it is probable that Rota's reflections were influenced by a famous report by a group of MIT scholars, emphasizing the limits of the current model of industrial development; see Meadows, Meadows, Randers, and Behrens (1972).

[10] We note that Rota interprets it as a sort of legacy of Jewish monotheism; see Rota (1991a), p. 30.

axiology, but above all on that of theory. This is a fundamental choice motivated by his adhesion to phenomenology. I must, then, say a few words about intentionality, which can be considered the fundamental phenomenological relation that "any good phenomenological vulgate" is summed up in the slogan: "every consciousness [...] is consciousness of something" (Spinicci, 2002, p. 94).

The meaning of this maxim, to which Rota himself has recourse,[11] is often misunderstood in a way that is particularly important for our purposes. The principal misunderstanding stems from the common and juridical sense of the term "intentionality," which contains

> the idea of a *tending*, of a *heading towards* whose goal is the object [...]. From these images [...] it is opportune to distance ourselves because, in general, it is far from true that every experience is animated by a tension towards a particular object.[12]

The validity of this theoretical remark can be grasped by means of simple considerations demonstrating how the tension towards an end is only *one* among the multiple modalities of our everyday experience. Each one of us can "in fact *look for* the house keys, but this does not mean that in doing so one simply *sees* the table, shelf, or bookcase where they might be."[13] This seeing is independent of the object looked for, as a simple variation demonstrates: if instead of the keys we were looking for a pen, in the same context we could still come up against the table, shelf, and bookcase.

Thus the *something* that is necessary to saturate the fundamental phenomenological relation cannot be identified with or reduced to an end by assimilating intentionality to teleology. On the contrary, the end as an object of consciousness represents nothing but a particular case of intentional relation and, therefore, only equivocally can an intrinsic finalism be attributed to phenomenology. This is a fundamental aspect that must be taken into account, also and above all by any phenomenologically-oriented axiology.

Now, I wish to delve further into the question of the "glance" in perceptive space, since it synthesizes numerous nonteleological aspects of phenomenological investigation and underscores the function of a number of invariant elements that are of great importance for phenomenology. It may in fact be said that, from the phenomenological standpoint, it is not the subjective structures that determine the regularities we observe in the universe but—on the contrary—it is the latter that determine the subject and the world as poles of the intentional relation (Rota, 1973b; see 1986a, p. 248).

Among these regularities, shading—which makes objects present themselves as partially visible—is of particular importance, as is shown by the laws of perspec-

---

[11] See Rota (1974b).

[12] Spinicci (2002), p. 94 (emphasis in original) .

[13] Loc. cit. (emphasis in original).

tive in perceptive space.[14] In fact the shifting of the viewpoint makes it possible to accede to parts of an object that were not visible before, on the condition that the previously focalized parts vanish. Analogously, consciousness can be assimilated to a sort of intentional horizon that delimits and divides a field of vision composed of backgrounds and foregrounds, of visible, shadow, and blind areas. This produces a distortion that, making it possible to generalize shading to all intentional phenomenon, allows us to clarify some aspects of Rota's thinking (Rota, 1973a; see 1986a, pp. 167–173).

Let us focus on the very concepts we have been engaged with here. At present, the foreground of the reader's attention is occupied by a phenomenological problem that represents the object that saturates the intentional structure. In the margins of this field further themes progressively shade off, while others are outside the horizon of present consciousness altogether.

Just as the movement of the eyeballs and the bending of the head allow the reader to see the objects located around or behind the book, so the shifting of his or her attention makes it possible to make other things explicit; for example, the concept of progress we have previously examined. In this way the content of these pages leaves the focus (in the optical sense) of the intentional field to enter more peripheral areas, until it is concealed in shadow areas that at the moment are invisible. This route (analogously to the shifting of the glance) is not reducible to an end, nor is it totally active, since other themes can interfere with the argument we attempt to focus on, as a lazy pupil—or a tired, or overly curious scholar—knows all too well.

One of the central elements of the generalization that permits us to pass from a perceived to an ideal object is represented precisely by the focus of the field of consciousness that constitutes a sort of foreground of attention. Going back to our parallel with the search for the house keys, we emphasize the fact that this focus (analogously to the glance) can encounter other objects that may be unimportant, or that may be significant. In this second case they capture our attention because they are intermediate stages of our original search, or because they are useful for other projects. They assert themselves in the intentional horizon even without having been deliberately sought after, because they have a force of their own.

As universal, the intentional structure also applies to mathematical entities that, like all objects, are grasped through shadings that, from one time to the next, bring out their particular aspects. This accounts for the importance of the different axiomatizations, proofs, and interpretations of the same mathematical object that permit the researcher to grasp its unsuspected aspects and properties. Rota suggests that "phenomenology can be considered as the explication of three ideas: as, already and beyond" (Rota, 1991a, p. 81). These unexpected aspects and properties express in the domain of mathematics the function of the "beyond" that

---

[14] See Husserl (1913), pp. 41–42.

Rota uses programmatically to characterize the—always necessary and never de-finitive—phenomenological overcoming of every perspective limit. In fact, doing research means "becoming aware of prejudices that stop us from seeing what is in front of us."[15]

4.3 ✧ Teleological Aspects of Mathematical Research

The aim of my analysis of nonteleological aspects has been to diminish the ele-ments of finality and design in the horizon of science, in order to understand their function better. In this regard we recall that Rota, in keeping with his antireduction-ist convictions, always speaks of "ends" and never "end" and describes the motor of mathematics in terms of a multiplicity of equally legitimate goals.[16] Rota in this respect is the heir of a liberal and pluralist tradition, fueled by the texts of Croce, which he articulates both in the epistemological and the axiological sense, justifying the reciprocal influence between mathematics and philosophy that characterizes his intellectual biography.

It is not fortuitous that Rota's research, centered on the finite and on the dis-crete, goes against the grain of most of 20th century mathematics, dedicated to the infinite and the continuous.[17] If we interpret multiplicity as the "corollary of a dis-crete vision of the real"[18] then Rota's antireductionism can be interpreted as a sort of philosophical parallel of his mathematical research. In fact, it can be maintained that "in the multiple [...] there is a sort of original polymorphism of reality [...], an original plurality of phenomena."[19]

Based on what we have said up to now, it is necessary to articulate this mul-tiplicity further. In the same way in which the end is only a particular case of the universal phenomenon of intentional relation, so technological applicability is only one of the possible senses of "end." Consider that the values represented by simplicity and beauty on the one hand, and specific problems on the other, can be equally important finalities for mathematical research.[20] Hence it is possible to un-derstand that an articulated axiology sustains—albeit implicitly—Rota's reflections on mathematical research.

In particular, Rota judges especially significant the confidential (or, more pre-cisely, *secret*)[21] list of problems that represents the motivation of every mathemati-

---

[15] Rota (1985a), p. 101.

[16] Ibid., p. 102.

[17] Lanciani (2005), p. 13.

[18] Ibid., p. 14.

[19] Loc. cit.

[20] See Rota (1997a), pp. 121–133, and Palombi (2004).

[21] In an article published towards the end of his life Rota lists ten unsolved problems he particu-larly has at heart, even though they are apparently far from his fields of research. He justifies this contradiction by saying that "the closer to our heart a mathematical theory lies, the harder it is to talk about it" (Rota, 1998c, p. 45).

cian's work, and on which any new concept or sector is systematically put to the test. Rota describes this way of working as a full and proper heuristic method, which, *in primis*, leads the mathematician to sift the fruits of this constant effort[22] through a rigorous process of selection based on logic and proof, in order to eliminate all the "crackpot ideas" that inevitably arise.[23] Among the results of this complex process (sustained both by the will and by reason) there are many theorems that were later judged to be brilliant.[24]

This is a phenomenon that has at times been synthesized under the exotic term "serendipity," as "sagacious observation [...] of one who [...] *observes without making conclusions on the truth content of what one sees*, with a suspension of judgment" (Garella, 2006, p. 29, emphasis in original). It is a dynamic that involves both a sort of phenomenological *epoché* and a kind of combinatorics of the possible cases. Both factors are particularly important in Rota's investigation.

This esoteric aspect of the work of mathematicians could constitute a partial justification for that sort of heterogenesis of ends thanks to which many intermediate stages necessary to prove a theorem or to conceive a new sector of mathematics are the result—not of a comprehensive and conscious effort but, rather, are assembled *a posteriori*.

To describe this complex historical dynamic we can find no better comparison than the one with the discovery of America, the fruit of a completely different project, which intended to find a new and shorter route to the Indies. It cannot be denied that the undertaking was pursued with systematicity and rationality; but it must also be said that the new continent was found in an entirely unexpected way. These considerations show why it is misleading to believe that

> mathematics, and scientific problems more generally, are formulated and solved in response to practical necessities [...]. Apart from science fiction novels, events of this type occur very rarely [...]. What does occur is very different. When a problem of application arises, the technologists throw themselves upon the scientific literature and go hunting for something that can be of help [...] usually [...] developed elsewhere for totally different reasons, or for no reason at all. (Rota, 1985e, p. 136)

Rota, in support of his thesis, draws up an interesting list of historic results, achieved with no regard for immediate technological objectives—such things as the discovery of the fast Fourier transform, of cluster analysis, of formal grammars, and of complex numbers (Rota, 1985e, p.137). But, in my opinion, the story of Professor Smith of the University of Terranova—a pseudonym behind which Rota conceals some fundamental details—is even more interesting (Rota, 1985e,

---

[22] Rota (1985a), p. 97.
[23] Ibid., p. 96.
[24] Ibid., p. 97.

pp. 135–136).[25] Professor Smith is looking for algorithms to subdivide rectangles into smaller ones, and publishes the results of this—apparently marginal—sector of research in serious but second-rate journals. The importance of his work changes dramatically with the advent of the microelectronic revolution, with chips produced on silicon wafers thanks to rectangular jigs. At that moment the algorithms for constructing rectangles are transformed from a sort of mathematical game of little importance, justified solely by the value of freedom of research, into an industrial tool of such great economic value that IBM technicians immediately set out "on a pilgrimage to Terranova."[26] For Rota the moral of this story is that

> the glory of mathematics [...] lies [...] in the exciting and wonderful fact that a theory developed to tackle a certain type of question often turns out to be the only way of solving entirely different problems, far from the ones for which the theory was conceived. These coincidences happen so frequently that they must necessarily belong to the deep essence of mathematics, and no philosophy of mathematics can avoid explaining them.[27]

4.4 ✧ The Indirect

I am convinced that Rota's challenge represents an important contribution to the epistemology of mathematics. For this reason, in my brief, programmatic conclusions, I shall seek to valorize it.

We have seen how mathematics imposes "lateral thinking"[28] on the researcher since, as Rota—following George Polya—affirms, "no mathematical problem is ever solved directly" (Rota, 1985a, p. 97; see Polya, 1962, p. 132). Our question, now, is the following: what is the general philosophical meaning of this *indirect*? First, it presupposes a unitary conception of mathematical reality of a Husserlian stamp (Husserl, 1959, Appendix VI, *The Origin of Geometry*). This is indispensable if the research of every scholar, developed from different starting points and different directions, is not to be transformed into solipsistic activity. A philosophic interpretation of this sort is endowed with indubitable heuristic value as is shown, for example, by Langlands' program that, proposing "a series of conjectures on the possible connections between disparate areas" of mathematics, represented

---

[25] Rota insisted that "this little story [...] is substantially authentic" (ibid., p. 136). For this reason it seems interesting to me that, according to the opinion of some friends and colleagues of Rota, this pseudonym would hide William Thomas Tutte (1917–2002), an Anglo-Canadian mathematician who was one of the greatest specialists in cryptography and graph theory of the last century.

[26] Rota (1985e) pp. 135–136

[27] Ibid., p. 22.

[28] I have extrapolated this expression from Du Sautoy (2003) to render an attitude that I previously referred to as a "sidelong glance."

the hinterland of the proof of Fermat's last theorem executed by Wiles (Odi-freddi, 2000a, pp. 27–28).

Second, emphasis on the indirect nature of mathematical proof subtends a distinctive context of discovery that it would be interesting to propose in the light of Bachelard's psychoanalysis of objective knowledge.[29] In this respect one might consider the proof of a theorem as a need that is primarily intellectual in nature. The satisfaction of this need stems from a logic of discovery described by Rota as the determination of the researcher, who unflaggingly tackles the problem from every possible direction. This self-discipline, on the one hand, possesses an autono-mous value of its own; on the other, it must be understood as the epiphenomenon of deep and continuous work, not controllable by the mathematician who slowly leads the discovery towards consciousness.[30]

This interpretation appears to repropose some aspects of the general neuroti-cizing character of human culture theorized by Freud.[31] In fact we are confronted with the impossibility of directly satisfying a need whose satisfaction is conditioned by the capacity to wait and to carry out complex and articulated strategies. The capacity to defer the satisfaction of a need for a long time, to satisfy it indirectly or substitutively, represents a reserve of sense and energy that is specifically charac-teristic of human culture.

At this point the perniciousness of the strictly economistic interpretation of science we noted at the beginning of this chapter has been made clear. The claim to be able to force scientific research into an increasingly narrow bureaucratic struc-ture in order to reduce it completely to a planning constituted by objectives, dead-lines, and hierarchies is counterproductive.

Rota's professional experience was that of a mathematician involved in the aca-demic, governmental, and military management of research in the United States; hence he can hardly be accused of humanistic ingenuousness. His defense of the "indirect" and his criticisms of a widespread interpretation of mathematics bear in mind not only the freedom of research but also the development of technology, the world of value, and that of practice.

---

[29] As a matter of fact, Bachelard limited himself to expressing the need for this undertaking and, rhapsodically, to drawing up an interesting list of historical cases, rather than developing a detailed analysis of a scientific theory, in reference to a specific psychoanalytic doctrine. See Bachelard (1938).

[30] This perspective is different from the one we have pursued here, following Rota. While I cannot go into the intricate problem of the relations between phenomenology and psychoanalysis on this occasion, I do wish to point out a distinction of great importance for our purposes: namely, that the former takes the form of an investigation of consciousness, of which the unconscious represents the edge, while the latter explains consciousness against the background of the un-conscious. I refer the reader interested in a psychoanalytic reading of science to Palombi (2003).

[31] See Freud (1929–1930).

Endnotes

[i] Vito Volterra (1860–1940), professor of mathematical physics, was one of the founders of the general theory of integral equations. His work in this field has been utilized recently in the quantitative models of ecology that study the relationship between predator and prey.

[ii] Enrico Fermi (1901–1954) was awarded the Nobel Prize for Physics for his research on slow neutrons. Beginning in 1927, when he became Professor of Theoretical Physics at Rome University, Fermi, together with Rasetti, brought into the Institute of Physics a group of such scientists as Edoardo Amaldi (1908–1989), Segrè, and Pontecorvo. The Via Panisperna Institute rapidly won international fame. Due to the racial policy of the Fascist regime, Fermi moved to the United States in 1939, where he continued his research at Columbia University. In America he carried out the first controlled nuclear reaction, participated in the project for the construction of the first atomic bomb, and authored many important studies in theoretical physics.

[iii] Franco Rasetti (1901–2002) took part in the Fermi group's research on slow neutrons. After a number of study periods in America, driven by the policies of the Fascist regime, in 1939 he moved to Canada to chair the Physics Department of Laval University in Quebec. In the following years, unlike the illustrious exponents of the Via Panisperna group, he refused to participate in the nuclear war effort, finally deciding to distance himself from research on nuclear physics. In 1947 he took a position at Johns Hopkins University, where he shifted his attention toward geology and paleontology. In the 1960s he concentrated on botany, authoring an important work of classification of Alpine flora. He returned to Rome in 1967 and then in 1977 moved to Belgium, his wife's native land, continuing his work as a naturalist. Rasetti's decision to devote himself to paleontology, geology, and botany stemmed from his desire to find fields of research that were free of military and government conditioning.

[iv] Bruno Pontecorvo (1913–1993), older brother of the famous film director Gillo, took his degree in physics in Rome in 1934 and became a member of the Via Panisperna group. In 1936 he moved to Paris and, subsequently, to the United States, where he participated in Anglo–Canadian war programs. In 1948 he went on to the Harwell center for nuclear research in Great Britain, and then emigrated to the Soviet Union in 1950. In the USSR, at the Nuclear Institute in Dubna, he took part in the research on elementary particles and on neutron physics.

ᵛ Emilio Segrè (1905–1989) took a degree in physics in Rome in 1928, and in 1935 became Professor of Experimental Physics at the University of Palermo. In 1938 the Fascist racial laws drove him to emigrate to the United States, where he lived for the rest of his life. During World War II he participated, together with Fermi, in the Manhattan Project that, in the Los Alamos laboratories, constructed the first atom bomb. After the war he concentrated on nuclear and elementary particle physics, and was awarded the Nobel Prize for Physics in 1959.

ᵛⁱ William Feller (1906–1970) was born in Zagreb, Croatia. He studied at the Universities of Zagreb and of Göttingen and began working at the University of Kiel, from which he had to flee in 1933 on account of the persecution of Jews. He moved first to the University of Copenhagen in 1934, and the following year to the University of Stockholm, where important research on probability was being done. In this period he worked principally as a consultant to scholars in other branches of science—statistics, economics, and biology—in which he was engaged with Probability Theory, which was not yet recognized and defined as a scientific discipline. Feller began to provide it with a theoretical structure. In 1939 he moved to the United States, where he taught at Brown University (1939–1945), Cornell University (1945–1950), and Princeton, developing his research in Probability Theory.

ᵛⁱⁱ Alonzo Church (1903–1995) studied at Princeton, Harvard, and Göttingen. He was professor of mathematics at Princeton from 1929 to 1967, and in the late 1960s he became professor of philosophy at the University of California, Los Angeles. He is remembered for his contributions to logic (in this field some of his works develop Gödel's research) and to informatics. In 1935 Church founded the Journal of Symbolic Logic. One of his 31 doctorate students was Alan Turing (1912–1954), the English mathematician expert in computability, who broke the cryptographic codes used by the Germans during World War II and devised the famous test that bears his name, utilized in the sphere of artificial intelligence.

ᵛⁱⁱⁱ Emil Artin (1898–1962) was born in the Bohemian city of Reichenberg, then part of the Austro-Hungarian empire (today Liberec in the Czech Republic), and studied at the Universities of Vienna, Göttingen, and Leipzig. He began his academic career at the University of Hamburg where he became full professor in 1926, and where he obtained important results in the theories of groups, semigroups, and topology, and solved one of the famous problems proposed by Hilbert in 1900. In 1929 he married one of his students, Natalie Jasny, a Jewess, and as a result he had to leave Germany when Hitler came to power. He emigrated to the United States in 1937 where he taught at Notre Dame (1937–1938), Indiana (1938–1946), and Princeton (1946–1958). In these two decades he concentrated on his teaching and on the writing of a number of texts that rapidly became classics. In 1958 he returned to his old teaching post at the University of Hamburg.

ⁱˣ Salomon Lefschetz (1884–1973) was a Russian Jew raised and educated in Paris. He emigrated to the United States in the early years of the last century to study at Clark University, where he took his PhD in 1911. He began to teach at Princeton in 1924, where he subsequently distinguished himself as chairman of the Mathematics Department until 1953. Lefschetz made major contributions to topology and to algebraic geometry.

ˣ John Kémeny (1926–1992) was a Hungarian Jew born in Budapest who moved to the United States in 1940. He enrolled at Princeton where he studied mathematics and philosophy, doing his bachelor's and his PhD under Church's supervision; in the same period he worked with Einstein as an assistant in mathematics. He immediately showed himself to be such a brilliant researcher that he took a year off from his studies to take part in the Manhattan Project at Los Alamos, working with Richard Feynman and John von Neumann. At Princeton he took his first teaching position as professor of philosophy but his successive academic career took place entirely at Dartmouth, where in 1953 he became a professor in the mathematics department, in 1955 chairman of the department, and finally, from 1970 to 1982, Rector of the University.

Moreover, in 1964 Kémeny (together with Thomas Kurtz) invented the language known as BASIC (Beginners' All–purpose Symbolic Instruction Code), and in 1979 was named by President Jimmy Carter to head the board of inquiry on the Three Mile Island nuclear power station incident.

[xi] Hermann Weyl (1885–1955) was born near Hamburg. He studied at the University of Göttingen where he attended the lectures of Hilbert, who influenced the first phase of his research, although he later distanced himself from his former teacher. In 1913 he was a professor at the University of Zurich where he was an associate of Einstein's; in 1930 he succeeded Hilbert and subsequently, due to the rise of the Nazi regime, he left Germany and moved to the United States, where he taught at Princeton until 1951. He divided the last part of his life between the United States and Switzerland.

His scientific research ranged over numerous fields of mathematics, from Lie algebra to number theory, differential and integral equations, mathematical physics, and differential geometry. From the philosophical viewpoint he was influenced by phenomenology, and in particular by Husserl's *Ideas Pertaining to a Pure Phenomenology* (1913).

[xii] Arthur Szathmary (b. 1916) studied at Harvard, where he took his bachelor's degree in 1937 and his PhD in 1941; he then taught at Princeton. His major works include *The Esthetic Theory of Bergson* (1937).

[xiii] John Rawls (1921–2002) studied first at Oxford and then at Princeton where, after taking his PhD, he taught in the 1950s. He became professor of philosophy at Harvard in 1962, a post he held for 28 years, finally becoming Professor Emeritus. One of his best-known works is *A Theory of Justice* (1971).

[xiv] Stanislaw Ulam (1909–1984), physicist and mathematician, was one of the most prestigious names of the Polish school of mathematics. In the second half of the 1930s, after losing much of his family to the Nazi persecutions, he emigrated to the United States where he decided to remain for the rest of his life, holding important academic, scientific, and government positions. His scientific contributions in fields ranging from pure mathematics to nuclear physics profoundly influenced the development of mathematics in its relation with science and technology.

[xv] Stefan Banach (1892–1945) was one of the most important exponents of the school of mathematics that flourished between the two world wars in the Polish city (now in Ukraine) of Leopoli. His fundamental contributions involved real variable functions, sets, functional analysis, and the theory of infinite-dimensional spaces.

[xvi] The mathematician and logician John von Neumann was born in Budapest in 1903, became a US citizen, and died in Washington in 1957. He studied at the Universities of Berlin, Budapest, and Zurich. In 1933 he became a member of the Institute for Advanced Study at Princeton and,

during the Second World War, worked on the atom bomb project at the Los Alamos National Laboratory. In 1955 he became a member of the U.S. Atomic Energy Commission. His many scientific contributions included the axiomatization of set theory; studies on game theory (together with Oskar Morgenstern); and research on probability theory, quantum mechanics, and numerical analysis.

[xvii] Kurt Gödel (1906–1978) studied at the University of Vienna. He also frequented the logical positivist Vienna Circle, but without sharing their general approach. He settled in the United States in 1939, working at the Princeton Institute for Advanced Studies. He is best known for the celebrated incompleteness theorem that bears his name. See Kline (1972), pp. 1203–1207, and Bottazzini (1990), pp. 415–425.

[xviii] Mark Kac (1914–1984) was educated at the University of Leopoli; he later moved to the United States where he taught at Cornell University, Rockefeller University, and the University of Southern California. He later held important political and institutional positions. He made important contributions in the field of probability theory and its applications.

[xix] Charles Méray (1835–1911) was one of the French theoreticians of the arithmetization of mathematics; see Kline (1972), pp. 1185–1186. Felix Klein (1849–1925) made important contributions to the study of non-Euclidean geometries and to topology; he is also remembered for an inaugural speech he delivered in 1872 on the occasion of his being named professor in Erlagen. See also Odifreddi (2000a), pp. 78–79.

Bibliography

This bibliography is divided into two sections: the first contains Gian-Carlo Rota's texts of philosophical character or interest; the second, the works of the other authors cited in the text, or used in my research.

After Rota's death, the books of his private library, conserved at his home on Massachusetts Avenue in Cambridge, were sold to the American Institute of Mathematics (AIM) in Berkeley. For its part, MIT collected in the Reading Room that bears Rota's name the material contained in his office and the books he donated to the university library. To this day I do not know the location of his vast personal archives (originally conserved at his home), which included notes, lecture texts, and correspondence. In my private archives I have catalogued chronologically in eight file-holders all the material Rota gave me, requesting that I conserve it and put it in order. This collection also includes some books Rota gave me as gifts and 25 audiocassettes with recordings of conferences and seminars. In the bibliography I indicate any unpublished material in my possession with the phrase "author's private archives."

The entire bibliography is compiled in alphabetical order by author; in the case of two or more works by the same author, the works are ordered chronologically, with a letter of identification to distinguish the works published in the same year. A few unpublished works of Rota's for which it was not possible to establish the date of composition are indicated as "undated" followed by a letter of identification. The title of articles and essays published in books, collective works, or journals are in quotes, and the works with two or more authors are arranged alphabetically on the basis of the name of the first author, except in the case of Rota where the coauthors are listed at the end.

The entries in the Rota bibliography are in general more complete than in the general bibliography, containing indications of the text's number of pages and, in some cases, the occasion for which it was originally written. The book reviews written by Rota are indicated with the term "Review" followed by the reference to the text reviewed. For unpublished material I have followed the same criteria indicating, wherever possible, the archives in which the original documents are conserved.

Only two anthologies of Rota's philosophical texts are readily available to the English-language reader: *Discrete Thoughts* (1986a) and *Indiscrete Thoughts* (1997a). In the footnotes, I cite first the original text referred to in the Rota bibliography followed—whenever the text is republished in *Discrete* or *Indiscrete Thoughts*—by (1986a) or (1997a) with the relative page number.

In the footnotes, page numbers of works for which the English translation is indicated in the general bibliography refer to the English-language edition. For example, in "Heidegger (1927), p. 100," the page number refers not to *Sein und Zeit* (1927) but to the English translation, *Being and Time* (1962).

✧ Publications by Gian-Carlo Rota

(1953). *On the solubility of linear equations in topological vector spaces.* Princeton, NJ: Princeton University.

(1956). *Extension theory of differential operators.* New Haven, CT: Yale University.

(1964). On the foundations of combinatorial theory I. Theory of möbius functions. *Wahrscheinlichkeitstheorie*, 2, 340–368.

(1969). Combinatorial analysis. In Bolhen (Ed.), *The mathematical sciences: A collection of essays* (197–208). Cambridge, MA: MIT Press. Republished in Rota (1986a) (49–62). Translated as "Analisi combinatoria" in the Italian translation of Bolhen (1969) (226–238). Republished in Rota (1993a) (52–67).

(1970). On the combinatorics of the Euler characteristic. In Mirsky (Ed.), *Studies in mathematics* (221–233). New York, NY: Academic Press. Republished in Rota (1995) (32–44).

(1973a). Edmund Husserl and the reform of logic. *Selected Studies in Phenomenology and Existential Philosophy*, 4, 209–305. Republished in Rota (1986a) (167–173). Translated in Italian as "Husserl e la riforma della logica" in Rota (1993a) (109–115).

(1973b). Grandeur et servitude du concept. *Cahiers Internationaux de Symbolisme*, 22–23, 79–83. Republished as (1977) "Controversial matters. A Heidegger prospectus." Phenomenology Information Bulletin, 1, 21–26, and "Heidegger" in Rota (1986a) (247–252). Translated in Italian as "Heidegger." Linea d'ombra, 24, 56–57. Republished in Rota (1993a) (116–121).

(1974a). *The end of objectivity. The technology and culture seminar at MIT.* Cambridge, MA: The MIT Press.

(1974b). A Husserl prospectus. *The Occasional Review*, 2, 99–106. Republished as "Husserl" in Rota (1986a) (175–181). Italian translation in Rota (1993a) (102–108).

(1974c). [Review of Kline (1972)]. *Bulletin of the American Mathematical Society*, LXXX(5), 805–807. Republished as "Mathematics and its history" in Rota (1986a) (157–161). Translated in Italian as "La matematica e la sua storia" in Rota (1993a) (47–51).

(1975). [Review of Reid (1970)]. *Philosophia Matematica*, XII, 76–80. Republished as "Misreading the history of mathematics" in Rota (1986a) (231–234). Translated in Italian as "Una cattiva lettura della storia della matematica" in Rota (1993a) (93–97).

(1976a). [Review of Brouwer (1975)]. *Advances in Mathematics*, 20(2), 285.

(1976b). [Review of Richardson (1971)]. *Advances in Mathematics* 20(3), 415.

(1976c). [Review of Hofmann (1974)]. *Advances in Mathematics*, 20(3), 415.

(1976d). [Review of Peterson (1973)]. *Advances in Mathematics*, 21(3), 366.

(1976e). [Review of Gandhi (1974)]. *Advances in Mathematics*, 22(1), 129.

(1976f). [Review of Fang and Takayama (1974)]. *Advances in Mathematics*, 22(3), 387.

(1977a). The wonderful world of Uncle Stan. [Review of Ulam, 1976]. *Advances in Mathematics*, 25(1), 94–95. Republished in Rota (1986a) (235–237). Translated in Italian as "Il meraviglioso mondo dello zio Stan" in Rota (1993a) (87–89).

(1977b). [Review of Sklar (1974)]. *Advances in Mathematics*, 24(1), 99.

(1977c). [Review of Corman (1975)]. *Advances in Mathematics* 23(2), 213.

(1977d). [Review of Adjukoewicz (1974)]. *Advances in Mathematics*, 24(2), 206.

(1977e). [Review of Shafer (1976)]. *Advances in Mathematics* 24(3), 341.

(1977f). [Review of Wilson (1975)]. *Advances in Mathematics*, 25(2), 188.

(1977g). [Review of Hardwick (1971)]. *Advances in Mathematics*, 26(2), 223.

(1978a). Combinatorial structure of the faces of the n-cube. *Siam Journal of Applied Mathematics*, 35(4), 689–694.

(1978b). [Review of Stuhlmann-Laeisz (1976)]. *Advances in Mathematics*, 27(1), 93.

(1978b). [Review of Quine (1966)]. *Advances in Mathematics*, 27(1), 93.

(1978c). [Review of Fogelin (1976)]. *Advances in Mathematics*, 27(1), 189.

(1978d). [Review of Malcom (1963)]. *Advances in Mathematics*, 28(1), 87.

(1978e). [Review of Cooper (1975)]. *Advances in Mathematics*, 28(2), 178.

(1978f). [Review of Winston (1975)]. *Advances in Mathematics*, 28(2), 178.

(1979a). Combinatorial math. *Science*, 204, 44–45.

(1979b). [Review of Dieudonné (1978)]. *Advances in Mathematics*, 34(2), 185–194.

(1979c). Coalgebras and bialgebras in combinatorics. *Studies in Applied Mathematics*, 61, 93–139. Republished in Rota (1995) (290–336).

(1980a). Howlett, J and Metropolis, N. (Co-eds.). *A history of computing in the twentieth century* (2nd ed. 1985). New York, NY: Academic Press.

(1980b). Metropolis, N. (Coauthor). *Preface* in Rota (1980a), xv–xvii.

(1981). *Introduction* in Davis and Hersh (1981), 1–4.

(1982). Kung, J.P.S. (Coauthor). Probabilità. In *Enciclopedia del Novecento* (Vol. IV, pp. 552–571). Rome: Istituto Nazionale dell'Enciclopedia Italiana.

(1983a). Context and project. Lectures on phenomenology. Typewritten notes based on the transcript of the lectures in the autumn course of 1982. Massachusetts Institute of Technology, Cambridge, MA. 189 pp.

(1983b). [Review of Boltzmann (1974)]. *Advances in Mathematics*, 47(3), 326.

(1983c). [Review of Kline (1980)]. *Advances in Mathematics*, 48(1), 115.

(1984a). Ulam. *The Mathematical Intelligencer*, 6, 40–42. Republished in Rota (1986a) (239–241), and with the title "Stan Ulam" in Rota (1997a) (60–62). Funeral oration at Los Alamos, 14 May 1984. Translated in Italian as "Ulam" in Rota (1993a) (90–92).

(1984b). Braithwaite, K., Kerr, D. M., Metropolis, N, and Sharp, D. (Co-eds.). *Science, computers and the information onslaught.* New York, NY: Academic Press.

(1984c). Braithwaite, K., Kerr, D. M., Metropolis, N., and Sharp, D. (Coauthors). "Preface" in Rota (1984b), XI–XIII.

(1984d). Gleason, A. M., Goldin, G. A., Metropolis, N., and Sharp, D. (Coauthors). Can science education cope with the information onslaught. In Rota (1984b) (263–271).

(1984e). Remarks on the present of artificial intelligence. Typescript of conference held at Los Alamos National Laboratory, February 1984. 17 pp. Author's private archives.

(1984f). Interview with Rota and Sharp. Typescript of the original transcript of N. Cooper's interview with Rota at Los Alamos, May 1984. Author's private archives.

(1984g). [Review of Chargaff (1978)]. *Advances in Mathematics*, 54(5), 500–503.

(1985a). Sharp, D. (Coauthor). Mathematics, philosophy and artificial intelligence. *Los Alamos Science*, 12, 92–104. Translated in Italian as "Matematica, filosofia e intelligenza artificiale" in 1992 Lettera Pristem (Dossier), 5, 1–16. Republished in Rota (1993a) (153–177). Edition based on Rota (1984f).

(1985b). Metropolis, N. (coauthor). *Introductory essay* in Rota (1980a). (2nd ed.). Republished in Rota (1986a) (191–193).

(1985c). The barrier of meaning. *Letters in Mathematical Physics*, 10, 97–99. Republished as "In memoriam of Stan Ulam: the barrier of meaning." 1986 *Physica*, 22D, 1–3; *Notices of the American Mathematical Society*, 36(2), 141–143; and with the original title in Rota (1997a) (55–59).

(1985d). *Artificial intelligence today*. Typescript revised and enlarged by Rota (1984e), Stein, P. R., (Coauthor). 19 pp. Author's private archives.

(1985e). *Osservazioni sull'intelligenza artificiale*. Bologna: Clueb. Republished in 1986 *Bollettino U.M.I.*, 6, 1–12; 1987 *L'elettronica*, LXXIV, 1015–1019; 1990 *Le Scienze quaderni*, 56, 5–11; and as "I.A." in Rota (1993a) (133–145). Inaugural address given on the occasion of the 900th anniversary of Bologna University. Italian edition revised and enlarged by Rota (1984e) and (1985d).

(1985f). Edmund Husserl's first logical investigation: expression and meaning presented in contemporary English. Typescript, 77 pp. Author's private archives.

(1985g). Words spoken at the memorial service for Mark Kac. University of Southern California, Los Angeles, 11 January 1985, 2 pp.

(1986a). Kac, M. and Schwartz, J. T. (Coauthors). *Discrete thoughts. Essays on mathematics, science and philosophy*. (2nd ed. 1992). Boston, MA: Birkhäuser. Japanese translation, Morikitá (1995).

(1986b). *Preface* in Rota (1986a) (ix–x).

(1986c). *Discrete Thoughts*. In Rota (1986a) (1–3).

(1986d). Kant. In Rota (1986a), pp. 243. Translated in Italian as "Kant," in Rota (1993a) (98–101). Text originally written as preface to Gulyga (1981).

(1986e). More discrete thoughts. In Rota (1986a) (263–264).

(1986f). [Review of Husserl (1980)]. *Advances in Mathematics*, 59(1), 95.

(1986g). [Review of Ingarden (1965)]. *Advances in Mathematics*, 59(3), 302.

(1986h). [Review of Gabbay and Guenthner (1983)]. *Advances in Mathematics*, 60(2), 237.

(1986i). [Review of Dreyfus and Hall (1982)]. *Advances in Mathematics*, 60(3), 360.

(1986l). [Review of Passmore (1985)]. *Advances in Mathematics*, 60(3), 360.

(1987a). Tre sensi del discorso in Heidegger. *Montedison Progetto Cultura*, 8. Republished in Rota (1993a) (122–129), and in Rota (1999a) (31–38). The English version was published with the title "Three Senses of 'A is B' in Heidegger" in Rota (1997a) (188–191).

(1987b). The lost café. *Los Alamos Science*, 15, 23–32. Republished in Rota (1997a), 122–129.

(1987c). Conversations with Rota. in *Los Alamos Science*, 15, 300–312. Dialogue between Rota and S. Ulam, transcribed and edited by F. Ulam.

(1987d). [Review of Halmos (1985)]. *American Mathematical Monthly*, 94, 700–702.

(1988a). Fine hall in its golden age: remembrances of Princeton in the early fifties. In P. L. Duren (1988). Republished in Rota (1997a) (3–20). Translated in Italian as "Fine Hall nell'età dell'oro," in Rota (1993a) (68–86); partially republished with the title "Soggetti della matematica," in 1993 *Il Sole* 24 ore, 160, 24.

(1988b). Syntax, semantics and the problem of mathematical objects. *Philosophy of Science*, 55, 376–386. Republished in Rota (1997a) (151–157), Sharp, D. and Sokolowski, R. (Coauthors).

(1988c). Gli italiani del MIT. *Epoca*, 39(1960), 86–94.

(1988d). Com'è intelligente quella macchina. *Saecularia Nova*, Bologna: Università di Bologna, 56–59.

(1988e). Gian-Carlo Rota. Publications. Typewritten list, divided into two parts: the first lists books, texts edited by Rota, and interviews (3 pp.); the second, articles (17pp.). Author's private archives.

(1988f). [Review of Lingis (1986)]. *Advances in Mathematics*, 68(1), 85.

(1988g). [Review of Baynes, Bohman, and McCarthy (1987)]. *Advances in Mathematics*, 71(1), 130.

(1988h). [Review of Husserl (1987)] *Advances in Mathematics*, 71(1), 131.

(1988i) [Review of Regvald (1987)]. *Advances in Mathematics*, 72(1), 169.

(1989a). *Fundierung* as a logical concept. *The Monist*, LXXII(1). Republished in Rota (1997a) (172–181). Translated in Italian as "*Fundierung*" in Rota (1993a). Republished in Rota (1999a) (21–30).

(1989b). Husserl's third logical investigation: a contemporary reading. Typescript of the conference held in August 1989 at the annual meeting of the Husserl Circle, Fort Collins (Colorado).

(1989c). [Review of Pylyshyin (1984)]. *Advances in Mathematics*, 73(1), 147.

(1989d). [Review of Harvey (1986)]. *Advances in Mathematics*, 73(2), 264.

(1990a). The concept of mathematical truth. *Nuova civiltà delle macchine*, 4, 145–150. Republished in 1990 *Mathematical Scientist*, 15, 65–73; in 1991 *Review of Metaphysics*, 44, 483–494; and in Rota (1997a) (108–120), with the title "The Phenomenology of Mathematical Truth." Translated in Italian as "Il concetto di verità matematica," 1991 supplement of *L'informatore vigevanese*, 20. Republished in Rota (1993a) (16–28). The French translation, delivered as a conference at the Collège de France, December 16, 1989, was published with the title "Les ambiguïtés de la pensée mathématique" in the *Gazette des matematiciens*, September 1990, pp. 54–64.

(1990b). Mathematics and philosophy: the story of a misunderstanding. *Review of Metaphysics*, December 1990, 260–271. Republished with the title "The Pernicious Influence of Mathematics upon Philosophy," in 1991 *Synthese*, 88, 8415–8419, and in Rota (1997a) (89–103). Translated in Italian as "Matematica e filosofia: storia di un malinteso," in *Bollettino UMI*, 4A, 295–307. Republished with the title "La nefasta influenza della matematica sulla filosofia," in Rota (1993a) (29–43), and in Rota (1999a) (39–54). The Spanish translation was published with the title "Matematicas y filosofia: la historia de un malcintendido," in March 1991 Ciencia, 5–12. The French translation was published with the title "L'influence néfaste des mathématiques sur la philosophie," in 1994 *Conjonctures*, 19, 159–177.

(1990c). Closing words at the Bologna Congress. In Bernabei, O., Borromei, A., and Orlandi, C. (1990) (239–241). Republished with the title "Philosophy and Computer Science" in Rota (1997a) (104–107).

(1990d). Dissertazione del laureando. In Ageno et al. (1990), pp. 51–55.

(1990e). Mathematics and the task of phenomenology. Typescript, 8 pp. Author's private archives.

(1990f). [Review of Wood (1989)]. *Advances in Mathematics*, 83(1), 133.

(1990g) [Review of Redhead (1989)]. *Advances in Mathematics* 84(1), 135.

(1990h). Course Program 1990–1991. Massachusetts Institute of Technology Bulletin, September, 223d.

(1991a) *The end of objectivity. The legacy of phenomenology. Lectures at MIT*, Cambridge, MA: MIT Mathematics Department.

(1991b). Transcript of the interview by F. Palombi for the *Enciclopedia multimediale delle scienze filosofiche* in December 1991 in Naples, 41 pp. Author's private archives.

(1992a). Ten rules for the survival of a mathematics department. MAA Focus, 12(6). Republished with the title "Ten Lessons for the Survival of a Mathematics Department," in Rota (1997a) (204–208).

(1992b). Sullo stato della scienza. *Linea d'ombra*, 71, 76–77. Republished in Rota (1993a) (Co-author F. Palombi) (146–152). Text of the address delivered at the inauguration of Carlo Cattaneo University on November 21, 1991.

(1992c). Remarks about the mythology of personality. Typescript February 1992. Author's private archives.

(1992d). One hundred years of phenomenology. Typescript of the lecture delivered on March 12, 1992 at the MIT Symposium, 6 pp.

(1992e). [Review of Rodriguez-Consuegra (1991)]. *The Bulletin of Mathematics Books and Computer Software*, 2, 6.

(1992f). Handwritten letter. June 1992, 3 pp. Author's private archives.

(1993a). *Pensieri discreti*. F. Palombi (Ed.). Milan: Garzanti.

(1993b). *Prefazione*. In Rota (1993a) (11–12).

(1993c). Gian-Carlo Rota. Publications. Typewritten list probably compiled by a collaborator. The document contains numerous errors of spelling and punctuation, of chronological order, and is incomplete.

(1993d). Combinatorics today. Inaugural address delivered at the International Meeting in Algebraic Combinatorics, Florence, June 21, 1993.

(1994a). Colloquio. Transcript of the private seminar held by F. Palombi in Vigevano, July 1994, 46 pp. Author's private archives.

(1994b). Course Program 1994–1995. *Massachusetts Institute of Technology Bulletin*, September, 522.

(1995). *On combinatorics. Introductory papers and commentaries* P. S. Kung (Ed.). Boston, MA: Birkhäuser.

(1996). *Curriculum vitae*. Typescript, 5 pp. Author's private archives.

(1997a). *Indiscrete thoughts* F. Palombi (Ed.). Boston, MA: Birkäuser 280 pp.

(1997b). Light shadows. Yale in the early fifties. In Rota (1997a) (21–38). Originally written as the inaugural address at the Congress in honor of J. T. Schwartz, held at the Courant Institute (New York University), May 19, 1995.

(1997c). Combinatorics, representation theory and invariant theory. The story of ménage à trois. In Rota (1997a) (39–54). Text of the inaugural address delivered at the Combinatorics Congress in honor of Adriano Garsia, Taormina, July 1994.

(1997d). The phenomenology of mathematical beauty. In Rota (1997a) (121–133). Lecture at the Boston Colloquium for the Philosophy of Science, November 17, 1992.

(1997e). The phenomenology of mathematical proof. In Rota (1997a) (134–150). Translated in Italian as "Fenomenologia della dimostrazione matematica," in Rota (1999a) (55–74).

(1997f). The barber of Seville or the useless precaution. In Rota (1997a) (Coauthors Sharp, D. and Sokolowski, R.) (158–161). This essay was originally conceived as a letter of reply to Professor Spalt's criticisms of Rota (1988b).

(1997g). Kant and Husserl. In Rota (1997a) (162–171). Text of the conference held at the Eighth International Kant Congress, Memphis, TN, March 2, 1995.

(1997h). The primacy of identity. In Rota (1997a) (182–187). Originally written for the Congress of Phenomenology in honor of Robert Sokolowski's 60th birthday, held at the Catholic University of America, Washington DC, November 11, 1994.

(1997i). Ten lessons I wish I had been taught. In Rota (1997a) (195–203). Originally written for the Rotafest, the conference organized in his honor at the MIT Mathematics Department, April 20, 1996.

(1997l). A mathematician's gossip. In Rota (1997a) (209–234); miscellaneous passages and short reviews.

(1997m). Book reviews. In Rota (1997a) (235–258). Collection of book reviews, most of them published in *Advances in Mathematics* and in Rotas column "Advanced Book Reviews" in *The Bulletin of Mathematics Books*.

(1997n). Klain, D. A. (Coauthor). *Introduction to geometric probability*. Cambridge, UK: Cambridge University Press.

(1998a). Che cos' 'è' matematica? In A. Cadioli et al. (Eds.), (1998a). Republished in Rota (1999a) (93–106).

(1998b). Analisi combinatoria, teoria della rappresentazione, teoria degli invarianti: storia di un ménage à trois. In *Bollettino dei docenti di matematica*, no. 37. Italian edition rewritten and enlarged by Rota (1997c). Republished in Rota (1999a) (Coauthor O. M. D'Antona) (75–92).

(1998c). Ten mathematics problems I will never solve. in *DMV Mittellungen*, 2, 45–52.

(1998d). Ten remarks on Husserl and phenomenology. Typewritten text of the conference held at the MIT Provost's Seminar, May 1, 1998, 11 pp.

(1998e). (1998, May 31). Tra logici e filosofi il gioco della verità. *La Stampa*, interview by P. Odifreddi. Republished in the complete version with the title "Intervista a Gian-Carlo Rota" in Odifreddi (2000b) (257–262).

(1998f). Manuscript notes of a draft by F. Palombi (1999b).

(1999a). *Lezioni napoletane*. Palombi, F. (Ed.). Naples: La città del Sole.

(1999b). Il primato dell'identità. In Rota (1999a) (107–116). Italian edition rewritten and enlarged by Rota (1997h) (Coauthor D. Senato).

(1999c). Fenomenologia del software. In Rota (1999a) (Coauthor F. Palombi) (117–128).

(2000). Prefazione. In Odifreddi (2000a) (VII–XII).

(2005). *Phénoménologie discrète*. Écrits sur les mathématiques, la science et le langage, Beauvais: Association pour la promotion de la Phénoménologie.

(2007). Lectures on being and time 1998. In M. van Atten (Ed.) *The New Yearbook for Phenomenology and Phenomenological Philosophy*, (1–99).

(undated a). Sample styles for introduction to Stan Ulam's collected papers. Typescript, 5 pp. Author's private archives.

(undated b). Wilson, A. (Coauthor). Philosophy and Truth. Typescript, 14 pp. Author's private archives.

(undated c). Untitled. Typewritten notes, 2 pp. Author's private archives.

(undated d). Introduction to MacMahon. In *Collected Papers*, Vol. II, Typescript. Author's private archives.

(undated e). A conversation with Gian-Carlo Rota. Written by an unidentified secretary of the MIT Mathematics Department, 10 pp.

(undated f). Translations of selected poems by Antonio Machado. Typescript, 5 pp. Author's private archives.

(undated g). The restlessness of mathematics. Typescript, 18 pp. Author's private archives.

✧ General Bibliography

Adjukoewicz, J.W. (1974). *Pragmatic Logic*. Dordrecht: Reidel.

Ageno, M. et al. (1990). *Conferimento delle lauree honoris causa*. L'Aquila: Università degli Studi de L'Aquila.

Alain. (1956). Janséniste et Jésuite. *Propos*, I, 467. Paris: Galimard (Pléiade).

Althusser, L. and Balibar, E. (1965). *Lire le capital*. Paris: Librairie François Maspero; (1970). *Reading Capital*. London: NLB.

Amsterdamski, S. (1981). Riduzione. In *Enciclopedia* (Vol. 12, pp. 62–75). Turin: Einaudi.

Anonymous (1950). Un certamen de psicologia estudiando todas las tendencias de ella ofreció el Colegio Americano, El Comercio. (Quito, Ecuador), 11-05, p. 7.

Bachelard, G. (1938). *La formation de l'esprit scientifique*, Paris: Vrin; (2002). *The formation of the scientific mind*. Manchester: Clinamen Press.

Bachelard, G. (1972). *L'engagement rationaliste*. Paris: PUF. pp. 27–34.

Baynes, K., Bohman, J., and McCarthy, T. (Eds.). (1987). *After philosophy, end or transformation?* Cambridge, MA: MIT Press.

Bednarek, A. R. and Ulam, F. (Eds.) (1990). *Analogies between analogies. The mathematical reports of S.M. Ulam and his Los Alamos collaborators*. Berkeley, Los Angeles and Oxford: University of California Press.

Benjamin, W. (1955). *Schriften*. Frankfurt/M: Suhrkamp Verlag. (1996–2003). *Selected Writings*. 4 Vols. Cambridge, MA: Harvard University Press.

Bergson, H. (1934). *La pensée et le mouvant. Essais et conferénces*, Paris: Alcan; (1992). *The creative mind: An introduction to metaphysics*. New York, NY: Carol.

Bernabei, O., Borromei, A., and Orlandi, C. (Eds.). (1990). *The brain and intelligence, natural and artificial*. Bologna: L'inchiostroblu.

Biemel, W. (1959). Les phases décisives dans le développement de la philosophie de Husserl. In *Husserl, Cahiers de Royaumont, Philosophie*, 3, 63–71.

Bolhen, G. (Ed.). (1969). *The mathematical sciences: A collection of essays*. Cambridge, MA: MIT Press; (1973). Translated in Italian as *Le scienze matematiche*. Bologna: UMI.

Boltzmann, L. (1974). *Theoretical physics and philosophical problems: Selected writings*. Dordrecht: Reidel.

Bonicalzi, F. (Ed.). (1982). *La ragione cieca*. Milan: Jaca Book.

Bonicalzi, F. (1987). *Il costruttore d'automi. Descartes e le ragioni dell'anima*. Milan: Jaca Book.

Bonicalzi, F. (1997). *La ragione pentita e il soggetto della città scientifica*. In Canguilhem and Lecourt (1997), pp. 9–54.

Bonicalzi, F. (1998). *A tempo e luogo. L'infanzia e l'inconscio in Descartes*. Milan: Jaca Book.

Bossi Fredigotti, I. et al. (Eds.). (1994). *Mi riguarda*. Rome: e/o.

Bottazzini, U. (1990). *Storia della matematica moderna e contemporanea*. Turin: UTET.

Boyer, C. B. (1968). *A history of mathematics*. New York, NY: Wiley & Sons.

Brower, L. E. J. (1975). *Collected works. Philosophy and foundations of mathematics*. Amsterdam: North-Holland.

Bufalo, R., Cantarano, G., and Colonnello, P. (2010). Natura storia Società. *Studi in onore di Mario Alcaro*. Milano: Mimesis.

Cadioli, A. et al. (Eds.). (1998a). *Scrittura e libertà*. Milan: il Saggiatore.

Cadioli, A. et al. (Eds.). (1998b). *Introduzione*. In A. Cadioli et al. (Eds.) (1998a) (9–15).

Canguilhem, G. and Lecourt, D. (1997). *L'epistemologia di Gaston Bachelard*. New edition edited by F. Bonicalzi, Milan: Jaca Book.

Carnap, R. (1937). *The logical syntax of language*. New York, NY: Kegan Paul.

Carnap, R., Hahn, H., and Neurath, O. (1929). *Wissenschaftliche Weltauffassung. Der Wiener Kreis*. Vienna: Artur Wolf Verlag. The scientific conception of the world. In S. Sahotra (Ed.). (1996).

Cellucci, C. (2002a). *L'illusione di una filosofia specializzata*. In M. D'Agostino et al. (Eds.). (2002) (119–137).

Cellucci, C. (2002b). *Filosofia e matematica*. Rome and Bari: Laterza.

Cellucci, C. (2009). Indiscrete variations on Gian-Carlo Rota's themes. In Damiani, D'Antona, Marra, Palombi (Eds.) (2009) (211–228).

Cerasoli, M. (1999). Il fascino discreto di Gian-Carlo Rota. *Bollettino dei docenti di matematica*, 39, 9–24.

Chargaff, E. (1978). *Heraclitean fire*. New York, NY: New York University Press.

Cimatti, F. (2002) Il senso indaga i suoi confini. In *il Manifesto*, 15 marzo, 15.

Comte, I. A. (1851–1854). *Système de politique positive, ou traité de sociologie instituant la religion de l'humanité*. 4 Vols. Paris: Carilian-Goeury et Dalmont.

Conrotto, F. (2006). *Statuto epistemologico della psicoanalisi e metapsicologia*. Rome: Borla.

Cooper, D. E. (1975). *Knowledge of language*. New York, NY: Humanities Press.

Corman, J. W. (1975). *Perception, common sense and science*. New Haven, CT: Yale University Press.

Costa, V. (2002a). Costituzione e teoria del'esperienza. In Costa, Franzini, and Spinicci (2002) (158–183).

Costa, V. (2002b). Etica e storia. In Costa, Franzini, and Spinicci (2002) (213–232).

Costa, V. (2002c). Questioni limite: i manoscritti della ricerca. In Costa, Franzini, and Spinicci (2002) (233–244).

Costa, V. (2002d). Realtà, storia e mondo. in Costa, Franzini, and Spinicci (2002) (245–263).

Costa, V. (2002e). L'eresia di Heidegger. In Costa, Franzini, and Spinicci (2002) (264–277).

Costa, V. (2002f). Fenomenologia e filosofia analitica. In Costa, Franzini, and Spinicci (2002). (294–318).

Costa, V. (2010). *Fenomenologia dell'intersoggettività. Empatia, socialità, cultura.* Rome: Carocci.

Costa, V., Franzini, E., and Spinicci, P. (2002). *La fenomenologia.* Turin: Einaudi.

D'Agostini, F. (1997). *Analitici e continentali. Guida alla filosofia degli ultimi trent'anni.* Milan: Cortina.

D'Agostino, M. et al. (Eds.). (2002). *Logica e politica. Per Marco Mondadori.* Milan: Fondazione Mondadori-il Saggiatore.

D'Alessandro, P. (1980). *Darstellung e soggettività (Saggio su Althusser).* Florence: La Nuova Italia.

Dalmasso, G. (Ed.). (1990a). *La de-costruzione. Testualità e interpretazione.* Pisa: ETS.

Dalmasso, G. (1990b). Etica e conferimento del senso. In Dalmasso (1990a) (47–57).

Damiani, E., D'Antona, O., Marra, V., and Palombi, F. (Eds.). (2009). *From combinatorics to philosophy. The Legacy of G.–C. Rota.* New York, NY: Springer.

D'Antona, O. (1999a). Il mio ricordo. *Bollettino dei docenti di matematica* (Cantone Ticino), 39, December, 25–28.

D'Antona, O. (1999b). *Introduzione alla matematica discreta.* Milan: Apogeo.

Davis, P. and Hersh, R. (1981). *The mathematical experience.* Boston, MA: Birkhäuser.

De Broglie, L. (1954). Foreword. In the English edition of Duhem (1906).

Dennett, D. (1991). *Consciousness explained.* Boston, MA: Little, Brown.

Derrida, J. (1962). *Introduction à l'origine de la géometrie de Husserl.* Paris: Presses Universitaires de France; (1978). *Edmund Husserl's origin of geometry: An introduction.* Lincoln, NE: University of Nebraska Press.

Derrida, J. (1967a). *De la grammatologie.* Paris: Les Éditions de Minuit; (1974). *Of grammatology,* Baltimore, MD: Johns Hopkins University Press.

Derrida, J. (1967b). *La voix et le phénomène,* Paris: PUF. *Speech and phenomenon.* Evanston, IL: Northwestern University Press.

Dieudonné, J. (1978). *Abrégé d'histoire des mathématiques 1700–1990.* 2 Vols., Paris: Herman.

Dreyfus, H. L. and Hall, H. (Eds.). (1982). *Husserl, intentionality and cognitive science.* Cambridge, MA: MIT Press.

Duhem, P. (1906). *La théorie physique. Son objet et sa structure,* Paris: Chevalier et Rivière (2nd enlarged edition, 1914); (1991). *The aim and the structure of physical theory.* Princeton, NJ: Princeton University Press.

Dulio, R. (2000). Giovanni Rota. Un modernista a Vigevano. *Viglevanum,* March, 8–15.

Duren, P. L. (Ed.). (1988). *A century of mathematics in America.* Providence, RI: American Mathematical Society.

Du Sautoy, M. (2003). *The music of the primes.* New York, NY: Harper Collins.

Fang, J and Takayama, K. P. (1974). *Sociology of mathematics and mathematicians.* Hauppauge, NY: Paideia.

Feist, R. (2002). Weyl's appropriation of Husserl's and Poincaré's thought. *Synthese,* 132(3), 273–301.

Ferraris, M. (1988). *Storia dell'ermeneutica.* Milan: Bompiani.

Feyerabend, P. (1975). *Against method.* (Third ed. 1993). London: Verso.

Floistad, G. (Ed.) (1981). *Contemporary philosophy. A new survey.* Vol. I. The Hague: Nijhoff.

Fogelin, R. J. (1976). *Wittgenstein.* Boston, MA: Routledge & Kegan.

Fontana, M. (1996). *Percorsi calcolati. Le nuove avventure della matematica.* Recco (Genova): Le Mani.

Foucault, M. (1963). *Naissance de la clinique. Une archéologie du regard médical.* Paris: PUF; (1991). *The birth of the clinic: An archaeology of medical perception.* London: Routledge.

Foucault, M. (1969), *L'archéologie du savoir.* Paris: Gallimard; (1972). *The archaeology of knowledge.* New York, NY: Pantheon Books.

Franzini, E. (1991). *Fenomenologia. Introduzione tematica al pensiero di Husserl.* Milan: Franco Angeli.

Freud, S (1929–1930). *Das unbehagen in der kultur.* Vienna: Internationaliter Psychoanalytischer. Civilization and its discontents. In *Standard Edition of the Complete Psychological Works of Sigmund Freud.* 1953–1974, Vol. 21, 64–145. London: Hogarth Press.

Gabbay, D. and Guenthner, F. (1983). *Handbook of philosophical logic.* Dordrecht: Reidel.

Gandhi, R. (1974). *Presuppositions of human communication.* Delhi: Oxford University Press.

Garella, A. (2006). Serendipity. In Conrotto (2006) (25–44).

Gasché, R. (1986). *The Tain of the Mirror: Derrida and the Philosophy of Reflection.* Cambridge, MA: Harvard University Press.

Geymonat, L. (Ed.) (1972–1976). *Storia del pensiero filosofico e scientifico,* 9 Vols., Milan: Garzanti.

Gioberti, V. (1840). *Introduzione allo studio della filosofia.* 3 Vols., Brussels: Marcello Hayez.

Goldman, J. R. (1995). The Cambridge School of Combinatorics. In Rota (1995) (XVI–XVIII).

Gould, S. J. (1981). *The Mismeasure of Man.* New York, NY: Norton.

Graubard, S. R. (Ed.). (1988). *The artificial intelligence debate. False starts, real foundations.* Cambridge, MA: MIT Press.

Guerra, T. (1994). Respiri paralleli. In I. Bossi Fredigotti et al. (Eds.), (1994) (61–62).

Gulyga, A. V. (1981). *Immanuel Kant,* Frankfurt/M: Insel Verlag; (1984). *Immanuel Kant.* Boston: Birkhäuser.

Hadamard, J (1896). Sur la distribution des zéros de la fonction ζ (s) et ses conséquences arithmétiques. *Bulletin de la Société mathématique de France*, 14, 199–220.

Hadamard, J. (1945). *The psychology of invention in the mathematical field.* Princeton, NJ: Princeton University Press.

Halmos, P. R. (1985). *I want to be a mathematician: An automathography.* New York, NY: Springer Verlag.

Hardwick, C. S. (1971). *Language learning in Wittgenstein's later philosophy.* Paris and The Hague: Mouton.

Harvey, I. E. (1986). *Derrida and the economy of différence.* Bloomington, IN: Indiana University Press.

Heidegger, M. (1912). Neue forschung für logik. In *Literarische Rundschau für katolisches Deutschland,* 38, 465–472, 517–524, 565–570.

Heidegger, M. (1927). *Sein und Zeit.* Tübingen: Niemeyer Verlag; (1962). *Being and time.* New York, NY: Harper & Row.

Heidegger, M. (1929a). *Kant und das problem der Metaphysik.* Bonn: Cohen; (1997). *Kant and the problem of metaphysics.* Bloomington, IN: Indiana University Press.

Heidegger, M. (1929b). *Was ist metaphysik?* Bonn: Cohen; (1977). *What is metaphysics?* In *Martin Heidegger: Basic writings.* New York, NY: Harper & Row.

Heidegger, M. (1929c). *Vom wesen des grundes.* Halle: Niemeyer Verlag; (1969). *The essence of reason.* Evanston, IL: Northwestern University Press.

Heidegger, M. (1967). *Wegmarken.* Frankfurt/M: Klostermann; (1998). *Pathmarks.* Cambridge, UK: Cambridge University Press.

Heidegger, M. (1969). *Zur sache des denkens.* Tübingen: Niemeyer Verlag; (1972). *On time and being.* New York, NY: Harper & Row.

Heidegger, M. (1972). *Frühe schriften.* Frankfurt/M: Klostermann.

Heidegger, M. (1975). *Die grundprobleme der phänomenologie.* Frankfurt/M: Klostermann; (1982). *The basic problems of phenomenology* (revised ed.). Bloomington, IN: Indiana University Press.

Heidegger, M. and Husserl, E. (1962). *Phänomenologische psychologie.* The Hague: Nijhoff; (1977). *Phenomenological psychology.* The Hague: Nijhoff.

Heller, K. D. (1964). *Ernst Mach: Wegbereiter der modernen physik.* Vienna and New York, NY: Springer Verlag.

Hofmann, J. F. (1974). *Leibniz in Paris, 1672–1676; His growth to mathematical maturity.* Cambridge, UK: Cambridge University Press.

Husserl, E. (1891). *Philosophie der arithmetik. Psychologische und logische untersuchungen,* Halle: Pfeffer; (2003). *Philosophy of arithmetic.* Dordrecht: Kluver.

Husserl, E. (1900–1901), *Logische untersuchungen.* 2 Vols. Halle: Niemeyer; (1970). *Logical investigations.* 2 Vols. London: Routledge; (2001) (All references are to Volume II).

Husserl, E. (1913). *Ideen zu einer reinen phänomenologie und phänomenologischen philosophie. Allgemeine einführung in die reine phänomenologie.* Vol. 1. Halle: Niemeyer; (1982). *Ideas pertaining to a pure phenomenology and to a phenomenological philosophy —First Book: General introduction to a pure phenomenology.* The Hague: Nijhoff.

Husserl, E. (1929). *Formale und transzendentale logik.* Halle: Max Niemeyer; (1969). *Formal and transcendental logic.* The Hague: Nijhoff.

Husserl, E. (1950a). *Die idee der phänomenologie.* The Hague: Nijhoff; (1966). *The idea of phenomenology. Five lectures.* The Hague: Nijhoff.

Husserl, E. (1950b). *Cartesianische meditationen und Pariser vorträge.* The Hague: Nijhoff (Husserliana); (1960). *Cartesian meditations.* Dordrect: Kluwer.

Husserl, E. (1952a). *Ideen zu einer reinen phänomenologie und phänomenologischen philosophie. Phänomenologische untersuchungen zur konstitution.* Vol.2. The Hague: Nijhoff; (1989). *Ideas pertaining to a pure phenomenology and to a phenomenological philosophy—Second Book: Studies in the phenomenology of constitution.* Dordrecht: Kluwer.

Husserl, E. (1952b). *Ideen zu einer reinen phänomenologie und phänomenologischen hilosophie. Die phänomenologie und die fundamente der wissenschaften.* Vol.3. The Hague, Nijhoff; (1980). *Ideas pertaining to a pure phenomenology and to a phenomenological philosophy—Third Book: Phenomenology and the foundations of the sciences.* Dordrecht: Kluwer.

Husserl, E. (1956). *Erste philosophie.* The Hague: Nijhoff (Husserliana).

Husserl, E. (1959). *Die krisis der europäischen wissenschaften und die transzendentale phänomenologie.* The Hague: Nijhoff; (1970). *The crisis of european sciences and transcendental phenomenology.* Evanston, IL: Northwestern University Press.

Husserl, E. (1980). *Phantasie, bewusstsein, erinnerung.* The Hague: Nijhoff.

Husserl, E. (1987). *Vorlesungen über bedeutungslehre.* The Hague: Nijhoff.

Hyman, J. (1991). *Investigating psychology. Sciences of the mind after Wittgenstein.* London: Routledge.

Ingarden, R. (1965). *Der streit über die existenz der Welt. Formal ontologie.* Tubingen: Niemeyer.

Kac, M. (1985). *Enigmas of chance: an autobiography.* New York, NY: Harper & Row.

Kant, I. (1783). *Prolegomena zu einer jeden künftigen metaphysik die als wissenschaft wird antreten können.* Riga: Hartknoch; (1977). *Prolegomena to any future metaphysics.* Indianapolis, IN: Hackett.

Kline, M. (1972). *Mathematical thought from ancient to modern times.* (3 Vols.). Oxford: Oxford University Press.

Kline, M. (1980). *Mathematics, the loss of certainty.* Oxford: Oxford University Press.

Kuhn, T. S. (1962). *The structure of scientific revolutions.* Chicago, IL: University of Chicago Press. (Postscript added in 1970).

Lacan, J. (1966). *Écrits*. Paris: Editions du Seuil; (2001). *Écrits: A selection*. London: Routledge.

Lakatos, I. (1959–1961). What does a mathematical proof prove? Published posthumously in Lakatos (1978b).

Lakatos, I. (1962). Infinite regress and foundations of mathematics. *Aristotelian Society Supplementary*, 36, 155–184.

Lakatos, I. (1963–1976). *Proofs and refutations. The logic of mathematical discovery*. Cambridge, UK: Cambridge University Press.

Lakatos, I. (1978a). *The methodology of scientific research programmes. Philosophical papers*. Vol. I. J. Worrall and G. Currie, (Eds.). Cambridge, UK: Cambridge University Press.

Lakatos, I. (1978b). *Mathematics, science, and epistemology. Philosophical papers*. Vol. II. J. Worrall and G. Currie, (Eds.). Cambridge, UK: Cambridge University Press.

Lanciani, A. (2005). *Introduction. Gian-Carlo Rota: mathématicien et philosophe*. In Rota (2005) (7–25).

Lingis, A. (1986). *Phenomenological explanations*. The Hague: Nijhoff.

Losee, J. (1972). *A historical introduction to the philosophy of science*. (New edition 2001). Oxford: Oxford University Press.

Mach, E. (1883). *Die mechanik in ihrer entwickelung historisch-kritisch dargestellt*. Leipzig: Brockhaus; (1960). *The science of mechanics. A critical and historical account of its development*. La Salle: Open Court.

Mach, E. (1905). *Erkenntnis und irrtum. Skizzen zur psychologie der forschung*. Leipzig: Barth; (1976). *Knowledge and error*. Dordrecht, Holland: Reidel.

Malcom, N. (1963). *Knowledge and certainty*. Englewood Cliffs, NJ: Prentice-Hall.

Mangione, C. (1976). La logica nel ventesimo secolo. In Geymonat (Ed.) (1972–1976), Vol. VII, 193–401.

Marini, A. (Ed.). (1982a). *Martin Heidegger. Il senso dell'essere e la "svolta": antologia storico-sistematica del "primo Heidegger."* Scandicci (Florence): La Nuova Italia.

Marini, A. (1982b). Introduzione storico-sistematica. In Marini (Ed.) (1982a), pp. VII–XC.

Mastronardi, L. (1962). *Il maestro di Vigevano*. Turin: Einaudi.

Meadows, D. H., Meadows, D. L., Randers, J., and Behrens, W. H. (1972). *The limits to growth: A report for the club of Rome's project on the predicament of mankind*. New York, NY: New American Library.

Merleau-Ponty, M. (1945). *Phénoménologie de la perception*. Paris: Gallimard; (2002) *Phenomenology of perception*. London: Routledge.

Mirsky, L. (Ed.). (1970). *Studies in mathematics*. New York, NY: Academic Press.

Monk, R. (1990). *Ludwig Wittgenstein. The duty of genius.* London: Jonathan Cape.

Moriconi, E. (1981). Presentazione. In Weyl (1932). *Il mondo aperto.* (7–27). Turin: Boringhieri.

Motterlini, M. (2000). *Lakatos. Scienza, matematica, storia.* Milan: il Saggiatore.

Mugnai, M. (2009). Rota's philosophical insights. In E. Damiani, O. D'Antona, V. Marra, F. Palombi (Eds.). (2009) (241–249).

Nagel, T. (1994). Consciousness and Objective Reality. In R. Warner and T. Szubka (Eds.). *The Mind-Body Problem,* (63-68). Oxford: Blackwell.

Nasar, S. (1998). *A beautiful mind.* New York, NY: Touchstone.

Odifreddi, P. (2000a). *La matematica del Novecento.* Turin: Einaudi.

Odifreddi, P. (2000b). *Il computer di Dio. Pensieri di un matematico impertinente.* Milan: Cortina.

Oldroyd, D. (1986). *The arch of knowledge. An introductory study of the history of the philosophy and methodology of science.* London: Methuen.

Paci, E. (1965). *Funzione delle scienze e significato dell'uomo.* Milan: Il Saggiatore.

Palombi, F (1997). Epilogue. In Rota (1997a) (265–271).

Palombi, F. (1999a). La doppia vita di Gian-Carlo Rota. Riflessioni su fenomenologia e matematica. In Rota (1999a) (7–18).

Palombi, F. (1999b). Sulla natura della matematica. La ricerca di Gian-Carlo Rota tra filosofia e scienza. *Bollettino dei docenti di matematica* (Cantone Ticino), 39, December, 29–35.

Palombi, F. (2002). *Il legame instabile. Attualità del dibattito psicoanalisi-scienza.* Milan: Franco Angeli.

Palombi, F. (2003). Il rovescio dell'Università: psicoanalisi e filosofia nel XVII seminario di Lacan. *Bollettino filosofico.* Philosophy Department of the University of Calabria, 19, 359–376.

Palombi, F. (2004). La forma della mente matematica: contesto della scoperta e bellezza. *Ou. Riflessioni e provocazioni,* XV, 33–38.

Palombi, F. (2005). Une application mal interprétée: une contribution de G.-C. Rota à la philosophie des mathématiques. In Rota (2005) (129–139).

Palombi, F. (2009a) A minority view. Gian-Carlo Rota's phenomenological realism. In E. Damiani, O. D'Antona, V. Marra, F. Palombi (Eds.). (2009) (251–260).

Palombi, F. (2009b). *Jacques Lacan.* Rome: Carocci.

Palombi, F. (2010). *Il ritorno dello psicologismo: Rota e l'indagine fenomenologica sulle scienze.* In R. Bufalo, G. Cantarano, P. Colonnello (Eds.). (2010) (307–3160.

Pancaldi, G. (Ed.). (1999). *L'Europa delle idee. Cultura europea e tradizioni nazionali.* Bologna: Philosophy Department of Bologna University.

Passmore, J. (1985). *Recent philosophers*. London: Duckworth.

Penrose, R. (1989). *The emperor's new mind*. Oxford: Oxford University Press.

Peterson, P. L. (1973). *Concepts and language. An essay in generative semantics and the philosophy of language*. The Hague: Mouton.

Piaget, J. (1968). *Le structuralisme*. Paris: Presses Universitaires de France; (1970). *Structuralism*. New York, NY: Harper & Row.

Piana, G. (1977). Introduzione. *L'intero e la parte. Terza e quarta ricerca*. Milan: Il Saggiatore; a partial translation in Italian of Husserl (1900–1901), pp. 7–71.

Picardi, E. (1999). Analitici e continentali. In Pancaldi (Ed.). (1999) (67–72).

Poincaré, H. (1902). *La science et l'hypothèse*. Paris: Flammarion; Science and hypothesis. In Poincaré (2008) (9–199).

Poincaré, H. (1905), *La valeur de la science*. Paris: Flammarion; The value of science. In Poincaré (2008) (200–358).

Poincaré, H. (1908). *Science et méthode*. Paris: Flammarion; Science and methods. In Poincaré (2008) (359–546).

Poincaré, H. (2008). *The foundations of science: Science and hypothesis, the value of science, science and methods*. Charleston, SC: BiblioBazaar.

Polkinghorne, J. (1986). *One world: the interaction of science and theology*. London: SPCK.

Polya, G. (1962). *Mathematical Discovery*. New York, NY: Wiley.

Popper, K. (1991). Meccanismi contro invenzione creativa: brevi considerazioni su un problema aperto. In P. Strata et al. (Eds.). (1991) (7–18).

Pylyshyin, Z. W. (1984). *Computation and cognition*. Cambridge, MA: MIT Press.

Quine, W. V. (1966). *Ontological relativity*. New York, NY: Columbia University Press.

Radnitzky, G. (1979). Metodo. In *Enciclopedia del Novecento* (Vol. IV, pp. 202–218). Rome: Istituto Nazionale dell'Enciclopedia Italiana.

Rawls, J. (1971). *A theory of justice*. Cambridge, MA: Harvard University Press.

Redhead, M. (1989). *Incompletness, nonlocality and realism*. Oxford: Clarendon Press.

Regvald, R. (1987). *Heidegger et le problème du néant*. The Hague: Nijhoff.

Reichenbach, H. (1938). *Experience and prediction: An analysis of the foundations of science*. Chicago, IL: University of Chicago Press.

Reid, C. (1970). *Hilbert with an appreciation of Hilbert's mathematical work by H. Weil*. Berlin: Springer Verlag.

Richardson, J. A. (1971). *Modern art and scientific thought.* Urbana, IL: University of Illinois Press.

Ricoeur, P. (1981) Logique herméneitique. In G. Floistad (Ed.). (1981) (179–223).

Rodriguez-Consuegra, F. A. (1990). *The mathematical philosophy of Bertrand Russell: Origins and development.* Boston, MA: Birkhäuser.

Rorty, R. (1982). *Consequences of pragmatism.* Minneapolis, MN: University of Minnesota Press.

Rossi, P. (1995). *La filosofia.* 4 Vols. Turin: UTET.

Rota Flaiano, R. (1997a). Vigevano, Flaiano ed io. Interview in *La provincia pavese,* 24-10, 43.

Rota Flaiano, R. (1997b). Io in via Panisperna. Interview in *La provincia pavese,* 25-10, 40.

Rota Gasperoni, E. (1995). *Orage sur le lac.* Paris: l'école des loisirs.

Rota Gasperoni, E. (1996). *L'arbre de capulies.* Paris: l'école des loisirs.

Ryle, G. (1932). Phenomenology. In *Proceedings of the Aristotelian Society,* XI. Reprinted in Ryle (1990).

Ryle, G. (1949). *The concept of mind.* London: Hutchinson's University Library.

Ryle, G. (1954). *Dilemmas.* Cambridge, UK: Cambridge University Press.

Ryle, G. (1962). Phenomenology versus the concept of mind. In *Cahiers de Royaumont Philosophie,* 4. Reprinted in Ryle (1990).

Ryle, G. (1990). *Collected papers I. Critical essays.* Bristol: Thoemmes.

Sahotra, S. (Ed.). (1996). *The emergence of logical empiricism: From 1900 to the Vienna Circle.* New York, NY: Garland.

Sartre, J. P. (1943). *L'être et le néant.* Paris: Gallimard; (2003). *Being and nothingness.* London: Routledge.

Senato, D. (1999). Quando incontrai la matematica. In *Bollettino dei docenti di matematica,* 39, 37–40.

Shafer, G. (1976). *A mathematical theory of evidence.* Princeton, NJ: Princeton University Press.

Singh, S. (1997). *Fermat's last theorem: The story of a riddle that confounded the world's greatest minds for 358 years.* London: Fourth Estate.

Sini, C. (1965). *Introduzione alla fenomenologia come scienza.* Milan: Lampugnani Nigri.

Sini, C. (1987). *La fenomenologia e la filosofia dell' esperienza* (2nd ed. 1990). Milan: Unicopli.

Sklar, L. (1974). *Space, time and spacetime.* Berkeley, CA: University of California Press.

Sokolowski, R. (1964). *The formation of Husserl's concept of constitution.* The Hague: Nijhoff.

Sokolowski, R. (1988). Natural and artificial intelligence. *Daedalus,* 117(1). Republished in Graubard (Ed.). (1988) (45–64).

Sokolowski, R. (1992). *Pictures, quotations, and distinctions. Fourteen essays in phenomenology.* Notre Dame, IN: University of Notre Dame Press.

Sokolowski, R. (1997). Foreword. In Rota (1997) (XIII–XVII).

Sokolowski, R. (2000). *Introduction to phenomenology.* Cambridge, UK: Cambridge University Press.

Spinicci, P. (2002), Significato, intenzionalità e logica. In V. Costa, E. Franzini, and P. Spinicci (Eds.). (2002) (63–114).

Strata, P. et al. (Eds.). (1991). *L'automa spirituale. Menti, cervelli e computer.* Rome and Bari: Laterza.

Stuhlmann-Laeisz, R. (1976). *Kants logik: eine interpretation auf der Grundlage von Vorlesungen, veröffentlichten Werken und Nachlass.* Berlin: de Gruyter.

Szathmary, A. (1937). *The esthetic theory of Bergson.* Cambridge, MA: Harvard University Press.

Ulam, S. (1976). *Adventure of a mathematician.* New York, NY: Charles Scribner's Sons. Republished (1991), Los Angeles, CA: University of California Press.

Vanni Rovighi, S (1969). *La fenomenologia di Husserl: note introduttive. Appunti delle lezioni introduttive ai seminari per l'anno accademico 1968–69: pro manuscripto,* Milan, Celuc, Second edition 1988.

Varzi, A. (2004). Mereology. In E. N. Zalta (Ed.) (2004).

Vozza, M. (1990). *Rilevanze : epistemologia ed ermeneutica.* Rome: Laterza.

Weyl, H. (1932). *The open world.* New Haven, CT: Yale University Press.

Weyl, H. (1952). *Symmetry.* Princeton, NJ: Princeton University Press. 1982.

Wiles, A. (1995). Modular elliptic curves and Fermat's last theorem. *Annals of Mathematics,* 142, 443–551.

Wilson, D. (1975). *Presuppositions and non-truth-conditional semantics.* New York, NY: Academic Press.

Winograd, T. and Flores, F. (1986). *Understanding computers and cognition. A new foundation for design.* Norwood, NJ: Ablex.

Winston, P. H. (Ed.). (1975). *The psychology of computer vision.* New York, NY: McGraw-Hill.

Wittgenstein, L. (1922). *Tractatus logico-philosophicus.* London: Routledge & Kegan Paul.

Wittgenstein, L. (1953). *Philosophische untersuchungen.* Oxford: Basil Blackwell; (1958). *Logical investigations.* Englewood Cliffs, NJ: Prentice-Hall.

Wood, D. (1989). *The deconstruction of time.* Atlantic Highlands, NJ: Humanities Press.

Zalta, E. N. (Ed.). (2004). *Stanford encyclopedia of philosophy.* Stanford: Stanford University.

Zimonti, G. (1983). *Vigevano. Vicende storiche vigevanesi.* Vigevano: Corsico.

Index

Printed in the United States
by Baker & Taylor Publisher Services